Tourism, Indigeneity, and the Importance of Place

Anthropology of Tourism: Heritage, Mobility, and Society

Series Editor: Michael A. Di Giovine (West Chester University of Pennsylvania)

Mission Statement

The Anthropology of Tourism: Heritage, Mobility, and Society series provides anthropologists and others in the social sciences and humanities with cutting-edge and engaging research on the culture(s) of tourism. This series embraces anthropology's holistic and comprehensive approach to scholarship, and is sensitive to the complex diversity of human expression. Books in this series particularly examine tourism's relationship with cultural heritage and mobility and its impact on society. Contributions are transdisciplinary in nature, and either look at a particular country, region, or population, or take a more global approach. Including monographs and edited collections, this series is a valuable resource to scholars and students alike who are interested in the various manifestations of tourism and its role as the world's largest and fastest-growing source of socio-cultural and economic activity.

Recent Titles in the Series

Tourism, Indigeneity, and the Importance of Place: Fighting for Heritage at Australia's Last Frontier by Carsten Wergin

Re-Centering Women in Tourism: Anti-Colonial Feminist Studies, edited by Frances Julia Riemer

Transnational Yoga at Work: Spiritual Tourism and Its Blind Spots by Laurah E. Klepinger

Medical Tourism & Inequity in India: The Hyper-Commodification of Healthcare by Kristen Smith

A Sacred Vertigo: Pilgrimage and Tourism in Rocamadour, France by Deana L. Weibel

Women and Tourist Work in Jamaica: Seven Miles of Sandy Beach by A. Lynn Bolles

Encounters across Difference: Tourism and Overcoming Subalternity in India by Natalia Bloch

Study Abroad and the Quest for an Anti-Tourism Experience edited by John Bodinger de Uriarte and Michael A. Di Giovine

Tourism and Language in Vieques: An Ethnography of the Post-Navy Period by Luis Galanes Valldejuli

Tourism, Indigeneity, and the Importance of Place

Fighting for Heritage at Australia's Last Frontier

Carsten Wergin

LEXINGTON BOOKS
Lanham • Boulder • New York • London

Published by Lexington Books
An imprint of The Rowman & Littlefield Publishing Group, Inc.
4501 Forbes Boulevard, Suite 200, Lanham, Maryland 20706
www.rowman.com

86-90 Paul Street, London EC2A 4NE

British Library Cataloguing in Publication Information Available

Library of Congress Cataloging-in-Publication Data

Names: Wergin, Carsten, author.
Title: Tourism, indigeneity, and the importance of place : fighting for heritage at Australia's last frontier / Carsten Wergin.
Description: Lanham, Maryland : Lexington Books, [2024] | Series: Anthropology of tourism: heritage, mobility, and society | Includes bibliographical references and index. | Summary: "The book presents a long-term ethnographic study of arguably the largest environmental protest action in Australian history. Carsten Wergin offers a timely discussion of the sociocultural and political relevance of heritage and tourism for ecological preservation and the wider decolonial project in Australia and beyond"—Provided by publisher.
Identifiers: LCCN 2023041667 (print) | LCCN 2023041668 (ebook) | ISBN 9781793648259 (cloth) | ISBN 9781793648266 (ebook)
Subjects: LCSH: Heritage tourism—Australia. | Heritage tourism—Political aspects—Australia. | Environmentalism—Australia. | Ethnology—Australia.
Classification: LCC G155.A8 W47 2024 (print) | LCC G155.A8 (ebook) | DDC 919.9404—dc23/eng/20231013
LC record available at https://lccn.loc.gov/2023041667
LC ebook record available at https://lccn.loc.gov/2023041668

To Lina, Adelė, and Aurėja,
my most beloved fieldwork collaborators.

(To all I pledge an endless supply of Courthouse Market Mango Smoothies.)

Contents

List of Illustrations and Tables ix
Preface xi
Acknowledgments xv

1 Learning through Experience 1
2 'Nowhere Else But Here' 15
3 From Transculturality to Transecology 55
4 The Four Pillars of Settler-Colonialism 85
5 On Common Ground 123
6 Knowledge and Place-Making 147
7 Collaborative Science 179
8 All Heritage Is Collaborative 201

Appendix: Tourism and Mining Tables 205
Bibliography 223
Index 243
About the Author 249

List of Illustrations and Tables

FIGURES

Figure 1.1.	Map of the Lurujarri Heritage Trail	7
Figure 1.2.	The Industrial Precinct Proposed for Walmadany/James Price Point	10
Figure 1.3.	The Aboriginal Communities of Northwest Australia	11
Figure 2.1.	Trail Participants Rest in the Shade during Story Time	16
Figure 2.2.	Map of the Dampier Peninsula, Also Indicating Manari Rd Corner	25
Figure 2.3.	"Meet and Greet" Protest Camp at Manari Rd Corner	26
Figure 2.4.	Trail Participants Pass Tourists on Cable Beach	28
Figure 2.5.	The Eternal Fire at Walmadany/James Price Point	29
Figure 2.6.	We Love Broome Protest in the Church Yard	32
Figure 2.7.	Theresa and Phil Roe at Walmadany/James Price Point, 2013	43
Figure 2.8.	Float of the Filipino Community during Shinju Matsuri 2013	45
Figure 2.9.	Courthouse Markets, Broome	47
Figure 2.10.	Richard Hunter Telling about Country	51
Figure 4.1.	Protest Action on Mother's Day 2012 in Broome	99
Figure 4.2.	McAlpine Statue at Cable Beach, Broome	105
Figure 4.3.	McAlpine Oval in Broome Converted to Overflow Caravan Park	107
Figure 4.4.	Broome Reggae Band The Mexicans Setting Up at Pearl Luggers Café	108

Figure 4.5. Frans Hoogland and Binky 113
Figure 4.6. Sunset Yoga and Wedding at Cable Beach, Broome 119
Figure 5.1. Packing Up and Loading the Truck with Swags of Trail Participants 128
Figure 5.2. Trailers at the Old Hostel under the Tamarind Tree 130
Figure 5.3. Trail Participants at Walmadany/James Price Point 132
Figure 5.4. Skinning the Killer 140
Figure 6.1. The Altar at Beagle Bay Church 151
Figure 6.2. Goolarabooloo/Jabirr Jabirr Claim Group Meeting Behind Closed Doors 152
Figure 6.3. No Gas Campaign Meeting at Lotteries House 165
Figure 6.4. Drill Rig at Walmadany/James Price Point 169
Figure 6.5. Whale Watching Platform at Murdudun 171
Figure 6.6. Whale Migration Path as Projected by Woodside Consultants 172
Figure 7.1. My Survey Toolkit 181
Figure 7.2. Banner Near the Entrance to Protest Camp Walmadany 187
Figure 7.3. Welcome Event for the *Steve Irwin* and Its Crew 188
Figure 7.4. The Jila (Waterhole) at Walmadany/James Price Point 190
Figure 7.5. Fireplace 192
Figure 7.6. Close Up of Turtle Bone 193
Figure 7.7. Phillip Roe Using the Grinding Stone 193
Figure 7.8. Phillip Roe Using the Grinding Stone 194
Figure 7.9. View South Overlooking the Dunes System 195

TABLES

Table A1. Tourism Tables (National) 205
Table A2. Tourism Tables for Western Australia 207
Table A3. Tourism Tables for Northwest Australia (Kimberley and Pilbara) 208
Table A4. Tourism Tables for Broome/Dampier Peninsula 209
Table A5. Mining Tables (National) 211
Table A6. Mining Tables for Western Australia 214
Table A7. Mining Tables for Northwest Australia (Kimberley) 218
Table A8. Mining Tables for Broome/Dampier Peninsula 220

Preface

On 10 February 2014, less than a year after Woodside Petroleum Ltd. withdrew from building a Liquefied Natural Gas (LNG) facility at Walmadany/James Price Point, the leader of the opposition movement, *Goolarabooloo* Law Man Mr Roe, aged 48, suffered a fatal heart attack. Faced with austerity and the violent expansionism of neoliberal market economics, Mr Roe had turned to his family, community, and "living country" for help in successfully staging what can be considered the largest environmental protest movement in Australian history. He had learnt to rely on country from his grandfather, the late Paddy Roe (OAM), founder of the *Lurujarri Heritage Trail,* who for more than 30 years has led its participants over Indigenous land along the Indian Ocean coast.

On 01 May 2018, Federal Court Justice Anthony North delivered the *Bindunbur* and *Jabirr Jabirr/Ngumbarl* Native Title claims determination for the Middle Dampier Peninsula to the *Jabirr Jabirr, Nyul Nyul,* and *Ngumbarl* people. The *Goolarabooloo* were not given legal status to continue their cultural practices on this land, as the court did not recognize them as Traditional Owners.

While this decision has provoked much concern about the future of the *Lurujarri Heritage Trail,* the research that preceded this ethnography was done in an atmosphere of hope. The fight for 'living country' continues, and this book would not have come into being without the determination of the *Goolarabooloo* and their supporters to keep it alive—to embrace moments of crisis not as potential losses but as a means of making the otherwise possible.

This atmosphere of hope is what I hope also to transmit in my writing, in which I seek to present the *Lurujarri Heritage Trail* not as a further example of what falls prey to late-liberal market economics but as that which allows access to a Dreaming of a world beyond the resources boom.

I would like to thank all those who have supported me in the completion of this project, and first and foremost the Indigenous people and communities in and around Broome and on the Dampier Peninsula: the *Bardi, Goolarabooloo, Yawuru, Jabirr Jabirr, Myinirr/Djugun, Nyigina, Numbarl,* and *Karajarri*. I would also like to express my gratitude to the many interview partners and collaborators who spent their time and energy answering my questions, sometimes over many hours and in repeated sessions.

In Broome, I am forever indebted to the Roe and Hunter families, Theresa Roe, the late Mr Roe, Margie Cox, Richard Hunter, Phillip Roe, Ronnie Roe and Molly Pilkington, Frans Hoogland, Jeanné Brown, the late Debora Vincent, Ketrina Keeley, Marilyn Tabatznik, Robyn Wells, Martin Pritchard, Stephen Pigram, *The Mexicans* (Tonchi Macintosh, George Bishop, Bart Pigram and Mel Raye, Gavin Pigram and Tania Bury), Tessa Mossop, Tegan Mossop, Jules and Gerard, Christine Elsasser and Shane Hughes, Josh Sibosado, Malcolm Lindsay, Anne Poelina and Ian Perdrisat, Peter Yu, Howard Pedersen, Sarah Yu, Kevin Fong, Neil Gower, Patrick Dodson, Mark Jones, and Peter Tucker. I would also like to acknowledge *Yalagarr* Childcare Services, Broome Circle House, Maca's Takeaway, Diver's Tavern and the Roey.

I would like to express my great appreciation and heartfelt thanks to Stephen Muecke and Prudence Black who in countless ways have been inspirational for this work and fundamental in making Australia a second home for my family and me. Stephen has been the greatest advisor, mentor, and friend and I am forever grateful to him for sharing some of his wit and wisdom with me.

Apart from the warm welcome we received from the Mueckes and their children, they also introduced us to many more key figures in the Walmadany/James Price Point conflict of which I would like to mention just one couple, Kirsty Cockburn and George Negus. They were the first to point out that the journey into this research began so much earlier than when we first set foot on Broome pindan. With this in mind, I would like to further express my gratitude to Burkhard and Cornelia Schnepel for making this project come to life.

In Heidelberg, I have had the privilege of receiving invaluable advice from Christiane Brosius. Further support came from Jenny Oesterle, Corinna Erckenbrecht, Daniel Münster, Guido Sprenger, Annette Hornbacher, and Bo Sax. I would like to thank my language editor Kerrie Le Lievre and my student assistants, Geva Herlyn, Anna Stähle, and Sophie Gindele, for their help. Again internationally, I would like to acknowledge the inspiration I drew from personal discussions with Tess Lea, Anna Haebich, Karen Fisher, Sarah Maddison, Fiona Ruth Allon, and the late Bruno Latour and his AIME-team. I am grateful to Michael Di Giovine for including this book in his series and the invaluable comments he gave during the preparation process, which leaves me equally indebted to Megan White, Ryan Dradzynski and everyone at Lexington for their dedication and patience. Finally, I would like to thank all my friends and colleagues at the Heidelberg Centre for Transcultural Studies (HCTS) and the Social Policy Research Centre (SPRC) in Sydney, who have provided me with such stimulating work environments.

This work was supported by a Marie Skłodowska-Curie Fellowship allocated within the seventh Framework Programme of the European Commission (FP7-MC-IOF) [grant number 253633]. Further substantial support was provided by the German Research Council (German Excellence Initiative) that funded the Research Group "The Transcultural Heritage of Northwest Australia" at Ruprecht Karls University Heidelberg, which I was fortunate to lead from 2014 to 2019.

Heidelberg, August 20, 2023

Acknowledgments

Chapters 2 and 5 draw on material that I first published in the article "Dreamings Beyond 'Opportunity': The Collaborative Economics of an Aboriginal Heritage Trail" (*Journal of Cultural Economy* 9/5 (2016): 488–506). Parts of chapter 3 have contributed to the chapter "From Transculture to Transecology. Coming to Terms with Multispecies Conviviality in the Education for Sustainable Development" (heiEDUCATION Journal 9 (2023): 83–95). The sections on "Tourism and Mining" in Chapter 4 draw on material that has first been published in "Questions of Value: Tourism and the Resources Boom," the introduction to a special section of *Australian Humanities Review* 53 that I coedited with Stephen Muecke. Parts of chapter 6 have first been published in "Responsibility = Ownership? An Ethnographic Moment in Native Title" as part of the volume *Mabo's Cultural Legacy. History, Literature, Film and Cultural Practice in Contemporary Australia* (coedited by Geoff Rodoreda and Eva Bischoff), as well as the article "Collaborations of Biocultural Hope: Community Science Against Industrialisation in Northwest Australia" (*Ethnos: Journal of Anthropology* 83/3 (2018): 455–472). Material presented in chapter 7 has first been published in "The 'White Magic' of Modernity: Retracing Indigenous Environmental Knowledge in Settler-Colonialist Australia," a contribution to the volume *Environmental Transformations and Cultural Responses: Ontologies, Discourses and Practices in Oceania* (coedited by Eveline Dürr and Arno Pascht). I thank the publishers of the above for their permissions to reproduce these works here.

Chapter 1

Learning through Experience

Australia has a unique environmental history where two of the most important global fields of contemporary culture and economics have confronted each other for decades: heritage preservation and the resources industry. On the one hand, the heritage of Aboriginal Australia continues to astonish scientists, who keep producing more evidence for a living past that is at least 60,000 years old. On the other hand, the Australian resources industry is what the country has relied upon for over two centuries and what saw it through the 2008 Global Financial Crisis (GFC). It continues to govern the policies and practices of Australian late-liberal politics.

Australia is often referred to as the "land of opportunity."[1] But these two conflicting domains, heritage preservation and resource exploration, have turned into a "wicked problem" in Australia because they are fundamentally opposed to each other: Indigenous heritage is entangled with the well-being of the land and sea, of Indigenous "country." But the well-being of the nation-state relies on the large-scale exploration of its natural resources that sees Indigenous country turned into industrialized space (Wergin and Muecke 2012). This wicked problem is at the heart of a conflict that emerged in the early 2000s, reached its peak in 2013, and continues to reverberate until this day: the fight for Walmadany/James Price Point, a significant Aboriginal cultural site on the Indian Ocean coast that was chosen as the site for the construction of a $45 billion AUS Dollar (€35 billion EURO) liquefied natural gas facility (LNG).

This book takes this conflict as the starting point for an analysis of contemporary Australian society, its Settler-colonialist structures, and movements that seek to separate from those with the help of Aboriginal knowledge, Lore, and custom.[2] The argument put forward is that the conflict over Walmadany/James Price Point highlights how the conventional "opportunity dreaming" of late-liberal politics is challenged by an Australian Dreaming beyond the resources boom, by ideals of an intimate connection between people and land, informed by Indigenous knowledge and put into practice through lived experience and collaboration.

This claim is grounded in original ethnographic material that stems from long-term fieldwork conducted between 2011 and 2018, during which I engaged with this conflict over preservation versus exploration at Walmadany/James Price Point. My field site is the Dampier Peninsula in the Kimberley region of Western Australia (hereafter WA). Because of its remoteness—many places remain accessible to conventional travelers only via helicopter—the region is famously described in tourist advertisements as Australia's "last frontier." With about 50,000 inhabitants and 423,000 square kilometers of landmass (roughly the size of Germany and Austria combined), the Kimberley remains one of the most scarcely populated regions in the world. Apart from its magnificent scenery, it is recognized by the global tourism industry as holding significant heritage values. In addition, creolization processes produced by Aboriginal, European, and South-East Asian migration (notably Japanese pearl divers) reach back to the late nineteenth century and form a further basis of its transcultural heritage.

For the past fifty years and more, this heritage has been intensely contested due to repeated and continuing proposals for large-scale explorations of, among other things, coal, iron ore, mineral sands, uranium, bauxite, and shale gas that is to be extracted throughout the Kimberley using the controversial technology of fracking. Conflicts between resource industries and heritage preservation initiatives are ongoing and have activated new and situated debates about the future of the region and its environment.

Numerous NGOs and representatives of Indigenous and non-Indigenous communities seek to preserve the heritage of the Kimberley and challenge the multibillion dollar exploration initiatives of state governments and the resources industry. The resulting collaborations between Indigenous and non-Indigenous groups, environmentalists, and local community members are remarkable for various reasons, not least because they are made up of people who might previously have contested each other's authority to speak for or even be on "Aboriginal country."[3]

The LNG facility at Walmadany/James Price Point was meant to process and store natural gas extracted from the Browse Basin in the Indian Ocean, roughly 425 kilometers offshore, with a value estimated at about $250 billion AUS Dollar. Construction was to begin in 2009, but Walmadany/James Price Point holds documented environmental and cultural heritage of great significance: burial grounds, artifacts, and spiritual sites as much as endangered fauna and flora. The conflict that developed over it thus presents a key example of a situation in which controversies over mining, tourism, heritage preservation, and land rights converged and significant frictions arose. Alongside people, culture, and heritage, equally significant and more-than-human actors in this conflict were Aboriginal "country," land, and sea.

Country intervened by making sites and wildlife visible to researchers, for example through heavy rains and winds that exposed human remains. Meanwhile, citizen science projects produced remarkable findings on whale migration and bilbies, monsoonal vine thickets, turtle nesting, and immense dinosaur trackways. A crucial factor for the success of the opposition movement was that Indigenous Elders and custodians of the land made their knowledge available to their diverse non-Indigenous collaborators, environmentalists, media representatives, tourists, and travelers, in order for them to understand the particular value of the region. This shared knowledge has been practiced for over thirty years as part of the *Lurujarri Heritage Trail* and allowed Indigenous and non-Indigenous people to work together—for the land and toward the preservation of a living heritage.

The successful movement for the preservation of Walmadany/James Price Point is thus an example for how to think "Indigeneity" in collaborative terms and as a means to address one of the greatest contemporary challenges: the project of decolonization. Indigeneity has been a preferred means for scientists to describe a people's long-term sustainable interactions with land and water (Weaver 2016, 31). Its global relevance to the reassessment of environmental values has been observed by numerous scholars, for example, in regard to the Brazilian Amazon (Bolaños 2011), Alaska (Dombrowski 2007), and Australia (Beckett 2012; Tonkinson 2007).

The acknowledgment of "Indigeneity" adds to the decolonial project "prerogatives of first occupancy, assumptions of environmental stewardship and cultural resistance, and efforts by mainstream society to make restitution" (Weaver 2016, 29; see also Pulitano 2014). Along these lines, while collaborators in the opposition movement against the LNG precinct were defending country for different and sometimes conflicting reasons, the qualities of "Indigeneity" were of unifying significance to them. Being Indigenous was not determined through blood lineage or kinship alone but through the acknowledgment of values that recognized "country" as a living entity, and a shared responsibility to protect its heritage.

In the course of this book I make use of an Indigenous concept of what in Western societies has become known as "the Dreaming" in order to highlight this entanglement of people and land as understood in Aboriginal Lore and custom and as an underlying force in the successful protest against the LNG facility. The recognition of Aboriginal heritage values by both Indigenous and non-Indigenous protesters set the stage for a widespread and diversified awareness of the environmental and social impacts of the proposed industrial development. I will argue that this awareness provides for a decolonial Australian imaginary—what I call an Australian Dreaming post-resources boom.

The impact of the LNG facility would have been particularly significant for the iconic tourist town of Broome, 50 kilometers south of the proposed precinct. Broome is one of the regional hubs of the Kimberley. It is a tourist town of about 15,000 inhabitants that can grow to up to 45,000 people during the "tourist season" (April to October), especially when the "grey nomads" move in—Australian retirees who have traded their houses for recreational vehicles (RVs) in which they crisscross the country based on where they will find the most appealing weather conditions.

Broome's attractiveness for tourists is enhanced by beautiful scenery such as the world-famous Cable Beach. Meanwhile, the town is unique not only because of its regional setting as a "gateway to the Kimberley," but also because about 30 percent of its population identify themselves as Aboriginal.[4] Because of this, environmental and cultural awareness have traditionally fostered a lush and artistic lifestyle in the community, inspired by the particular landscape that surrounds the town and that continues to attract new residents.

Astrid B., former journalist for the daily newspaper *The West Australian*, was based in Broome during the Walmadany/James Price Point conflict and describes the attraction of the town and the effect the landscape had on her and other people as follows:

> Astrid B.: It's a very intangible thing (. . .) The physical, built environment is pretty ordinary in places (. . .) But I think it's the combination of the physical beauty of the natural environment, which is just stunning (. . .) If I take the car and I go down onto Cable Beach (. . .) to get away from everyone else (. . .) I feel like I'm on holiday (. . .) You can completely tune out the rest of the world. It's quite a unique thing to be able to do in your hometown (. . .) I get the same thing when I go out driving in the bush as well (. . .) I love that feeling. It's amazing—that kind of freedom. And it's clean. The air is clean and unpolluted, and you notice that when you go to cities—how different the air is. And I think there is people here who *do* stuff. For a small community, you've got a really creative force. You've got lots of people who write books, make music (. . .)
>
> Carsten Wergin: Do you think those people are here out of similar reasons? Because they got attached to the environment?
>
> AB: Definitely. I mean if you're an artist it's visually striking (. . .) The photographers here (. . .) do wedding after wedding after wedding to make lots of money, and then they pursue their art. I guess it has that sense of being

> remote enough from the rest of the world that it feels like an escape (. . .) Certainly, in my years of working here, I've heard so many times people say, "I came here intending to travel around Australia, but I never left" (. . .) I suppose you've got a lot of people who've had that similar experience, and so they've bonded in a way (. . .) I'm speaking as an outsider (. . .) The locals that grew up here would probably tell a very different story. They'd tell a story about, "This is my childhood home. This is where I go fishing." But they seem to certainly have a love for it. And to a certain degree, they're kind of happy to share. They must've seen so many people come and go over the years, but it still has a welcoming feeling about it. (Transcript 121109_FP)

These qualities of Broome and the Kimberley—of welcoming, sharing, and becoming part of Indigenous country—are the foundations of the collaborations that emerged in opposition to the industrialization of Walmadany/James Price Point. My ethnography describes three "spheres"—the scientific sphere, the knowledge sphere, and the economic sphere—in which these collaborations played out. Central to my analysis of these spheres is that their actors collaborate while at the same time acknowledging the existence of other world(view)s and related values, in particular those of the Indigenous custodians.

These values are linked to a "living heritage," a term proposed by the *Goolarabooloo* themselves and one that I discuss in full detail below. It links to heritage as described by historian, geographer, and founding father of the discipline of heritage studies, David Lowenthal (1998, X), "a profession of faith in a past tailored to present day purposes." In other words, heritage is what people hold on to despite the social and cultural transformations they are confronted with. In the context of this inquiry, this profession of faith led to diverse collaborations between people and people, as much as people, land and sea, 'living country.'

My interest in these collaborations was motivated by the fact that they seemed to be not implemented from above but derived from below: they were not the result of conventional striving for economic opportunities that seek to extract value from the land, but of an acknowledgment of the particularities of place (cf. Escobar 2001). Though certainly not free of tensions and ruptures, people collaboratively acknowledged the environment and the heritage it hosts as being unique, valuable in itself and thus worth protecting (chapter 2).

This acknowledgment is of fundamental importance because the shared faith in the (Indigenous) heritage of the region offered people an alternative means of engaging with the pressing issues they were confronted with in regard to its preservation and development. It allowed collaborators to depart from conventional calls for the recognition of a "Multicultural Australia" based on, as the then Prime Minister Turnbull formulated it in 2017, "shared values of freedom, democracy, the rule of law and equality of opportunity" (The Australian Government 2017). Instead of such opportunity dreaming, collaborators positioned these tropes of freedom, democracy, the rule of law and equality within a wider, what I will term "transecological," project that took the entanglement with multiple world(view)s seriously (chapter 3).

To put it differently, collaborators in the opposition movement acknowledged particular heritage values, which they felt should not be streamlined into the image of one multicultural Australia. The difficulty for them, as much as for me, was to engage in these collaborations without falling back into dualisms such as "preservation versus exploration," "Indigenous versus non-Indigenous," "nature versus culture." Indigeneity, as defined above, offered an escape route here. The term evokes a recognition of country, land, and sea that is ontologically different. To paraphrase the definition of heritage proposed by Lowenthal, an emphasis on Indigeneity allows for a profession of faith that reaches significantly further than what the knowledge of a past tailored to present-day purposes has to offer, a trait that can be seen in the way that the collaborators' faith in Indigenous heritage values positioned "living country" and its living heritage beyond the late-liberal mainstream (chapter 4).

Along these lines, the engagement with "living country" and its living heritage as it was proposed by the *Goolarabooloo* and their collaborators, whether through scientific assessment or as tourist/traveler, prevailed over that of government and industry because it was conceived as a profession of faith. The *Goolarabooloo* and their collaborators always knew that "living country" would prevail because it provided them with *liyan*: a good feeling, strength enough to continue their opposition, or, as they emphasized, to "never give up" (chapter 5).

Here, *Bugarrigarra*—to introduce the West Kimberley term for what is commonly referred to as The Dreaming—aligns with non-Indigenous notions of sustainability and development. With regard to conventional definitions of heritage, it also describes a profession of faith that is not concerned with either the past or the present, because in *Bugarrigarra* both are always already present in the here and now (chapter 6).

Indeed, such an Indigenous perspective on heritage allowed protesters to recognize their entanglement with the environment and its consequences for its past, present, and future. In this sense, heritage became *living heritage*—that is, not a past tailored to present-day purposes, but a past that actively contributes to the present and that transforms with it. Recognizing their partnership with the environment through collaborative means allowed for those engaged in the protest movement to embrace the past, by committing themselves to a common future (chapter 7).

This proposition is, on the one hand, based on my ethnographic material and, on the other, informed by theories that have emerged out of the still fairly new academic field of Environmental Humanities. The aim of most researchers active in this field is to overcome the dualisms considered foundational to academic and non-academic discourse, theories, and methodological approaches that have kept conventional environmental studies separate from the humanities and even parts of social sciences, most notably, and maybe also most surprisingly, social anthropology. My ethnography of the multifaceted protest for the preservation of Walmadany/James Price Point ties in with related efforts to overcome such dualisms. In it, I set out from a particular site, Walmadany/James Price Point, and a particular experience of it, the *Lurujarri Heritage Trail* (figure 1.1). For 82 kilometers this Trail follows parts of an Aboriginal Song Cycle along the Indian Ocean coast, called the Northern Tradition. The

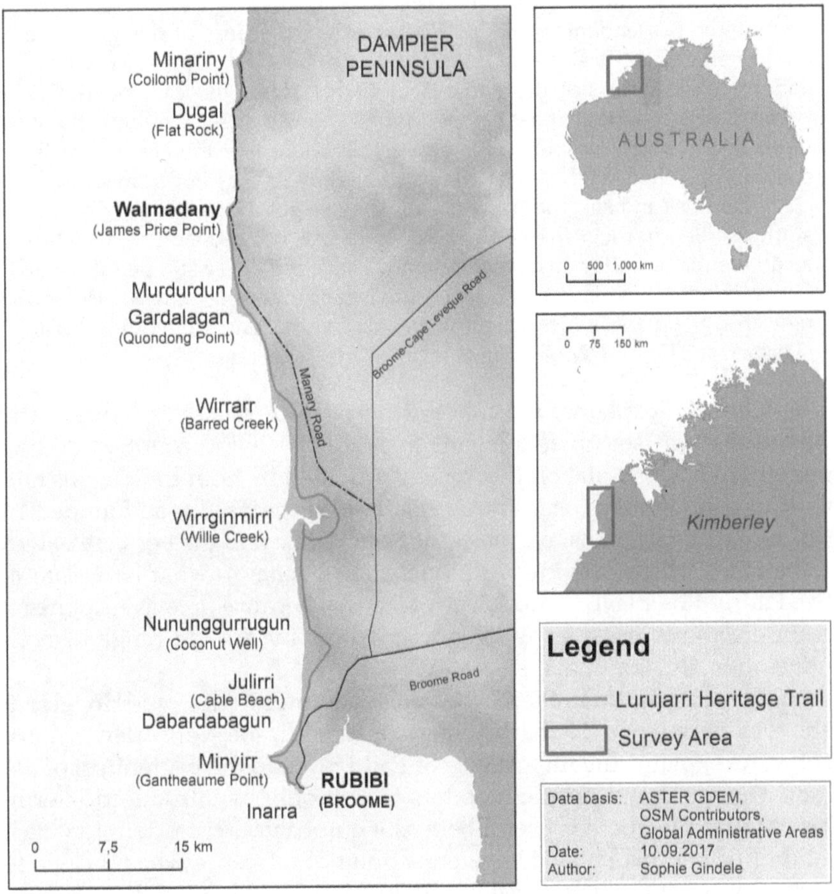

Figure 1.1. Map of the Lurujarri Heritage Trail. Photo: Carsten Wergin.

late Paddy Roe (OAM), senior Elder of the *Goolarabooloo* community who led Indigenous and non-Indigenous Trail participants across this land, founded it in 1987.

The Trail allows its participants to overcome dualisms because of the hybrid experience it offers, combined with the cultural hybridity of the *Goolarabooloo* community itself. Both successfully blur the boundaries between the Indigenous and the non-Indigenous. The interview excerpt below, from a conversation I had with Chris D., an Australian-born marine biologist in his mid-30s, who joined the *Goolarabooloo* in their protest movement, makes this hybridity explicit:

> *Goolarabooloo*, it's more recently defining the descendants of Paddy Roe. Obviously, the word *Goolarabooloo* was used before Paddy Roe came along. It meant "coming together of tribal culture" (. . .) An umbrella term to encapsulate all different cultural aspects that [Paddy Roe] knew. But (. . .) it really has become the descendants of Paddy Roe and that's because of this gas thing (. . .) Theresa [Paddy Roe's daughter] still calls herself a *Jabirr Jabirr* woman and there's people, not descendants of Paddy Roe, who call themselves *Goolarabooloo*. Although it has come to define descendants it shouldn't. But in terms of *Goolarabooloo*, in terms of culture, this whole gas campaign has really come to find them as these custodians (. . .) They're very hot-headed, especially the Roes and also the Hunters (. . .) They've got the core cultural values stuff, instilled in them from Paddy Roe, plus this hot-headedness, which is a bad combination for anyone coming in. That's why they stopped three or four mining developments. There was two sand mines (. . .) proposed and now this gas plant, which hasn't been stopped but is definitely in retreat. (Transcript 130513_1_ML)

As Chris D. explains, the *Goolarabooloo* are intrinsically hybrid. The community was set up as a means to preserve "living country" and its living heritage by collaborative means provided by both Indigenous and non-Indigenous members. The *Lurujarri Heritage Trail* was founded in 1987 as a further means to protect the Song Cycle and the heritage values of the land, which provide additional components to what is *Goolarabooloo*. Paddy Roe played a decisive role in this because he was recognized as the local authority on Indigenous matters by a wide range of non-Indigenous decision-makers.

Another key figure in the *Goolarabooloo* community is Frans Hoogland, who was born in the Netherlands and arrived in the Kimberley in 1968. Below, he explains the importance of Paddy Roe to the beginnings of the Trail. Though many people he refers to will only be introduced later in this inquiry, the quote is useful here as it emphasizes the extent to which Paddy Roe and his knowledge were a point of reference when it came to documenting or making decisions about Aboriginal land in the region:

> We had Richard Court come out here, Premier of Western Australia. He came and talked to us (. . .) We met that many different Premiers, Carmen Lawrence, Peter Dowling . . . a lot of them. We done the first Trail, then we talked to the sales department . . . [Anthropologist] Nick Green to do a management plan for it (. . .) Then those two ladies [Elizabeth Bradshaw and Rachel Fry] came up and we walked them. Old Man [Paddy Roe] came out by car, but basically we, them and us, walked the whole county from place to place, from midden to midden to note everything down (. . .) The report [Bradshaw and Fry 1989] came out, and at the same time, McAlpine was here. And he had a bit contact with Old Man and asked us to come for dinner . . . . So, we made him patron . . . . Saint patron of the Trail (. . .) After we made him patron, we got a truck out of him for the Trail. And we got money out of him to do more surveys (. . .)
>
> And the first two kilometers was non-development. I put it in [the report]. The first two kilometers you come from the ocean inland (. . .) you cannot do anything (. . .) No construction, no building, no nothing (. . .) And then, we had a buffer zone for low-key . . . Maybe tourism (. . .) When we came out with the report, we had the whole liberal shadow government coming here to interrogate the Old Man. They thought he was a big liar (. . .) Because all of a sudden, there was 78 new Aboriginal [burial] sites (. . .) I remember sitting down there with him. [Mr Roe][5] was not in the picture there. Not Richard [Hunter] (. . .) Ronnie [Roe] was [the three are Paddy Roe's grandchildren, *the author*] (. . .) They told them all: About the country, their names (. . .) All the old people he buried in the dunes (. . .) So that mob burned away. Then, we had some nice cartoons in the newspaper of white people living in the cave, and then the little boy asking, "Daddy, when can we get our house back?" (. . .)
>
> And then, TERREX sent mining all over the top of it. Over the two kilometers. But more of it up to four or five kilometers inland (. . .) So then, Nick Green came here and my older teachers at One Arm Point, and we all sit together to make the Song Cycle map (. . .) And with the song cycle, we went to court against TERREX (. . .) [Patrick] Dodson was helping us (. . .) then boss of KLC [Kimberley Land Council] (. . .) And before that case, whenever anybody wanted to do something on country they asked the Old Man, "What about this? How big is this place?" (. . .) They always were asking about places and sites. But now, the entire Song Cycle became one site (. . .) So it wasn't, "Oh, this is important and that is not important." It's all important! If you destroy this, then you destroy yourself . . . And we are still fighting about that one today. (Transcript 120807_4_FH)

Given the global situation of human-made environmental degradation, the importance of such collaborative efforts to preserve a "living country," fauna and flora, land and sea, needs no further justification. In addition, initiatives such as the *Lurujarri Heritage Trail* are sought-after means for people to partake in preservation, be it by doing a voluntary year of social service or by booking into Indigenous tourism experiences to learn more about sustainable, entangled livelihoods.

However, prior to 1987 when the Trail was inaugurated, the values of Indigenous culture and heritage to the promotion of what have become lucrative tourism experiences were largely peripheral to economic thinking. Back then, the land over which the Trail leads had repeatedly been suggested for sand mining. In those days, the core *Goolarabooloo* families—the Roe family and the Hunter family, which descend from two daughters of Paddy Roe—had staged successful protest actions against these industrialization proposals. What Frans Hoogland refers to at the end of the interview excerpt above is Walmadany/James Price Point as both an integral part of the Song Cycle and the last in a long series of conflicts with the resource industry.

The latest proposal by Australia's largest independent oil and gas company, Woodside Petroleum Ltd. (hereafter Woodside), would have brought an additional 6,000–8,000 workers to the region. It would also have included the construction of a large port facility with an estimated two-kilometre jetty and significant additional ship movements (figure 1.2).

I accompanied the multifaceted opposition movement to this proposal for almost two years *in situ*, until in April 2013 Woodside officially withdrew from building the facility. Artists, scientists, lawyers, politicians, environmental activists, Indigenous groups and local community members successfully counteracted the industrialization proposal. Their actions delayed the beginning of construction until, further supported

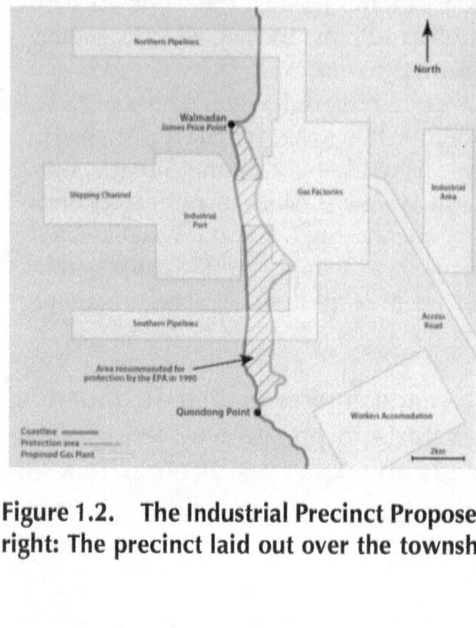

Figure 1.2. The Industrial Precinct Proposed for Walmadany/James Price Point. On the right: The precinct laid out over the township of Broome. *Source*: Environs Kimberley.

by a downturn in oil and gas prices, Woodside was forced to declare its proposal no longer viable.

This opposition literally found its "common ground" in the *Lurujarri Heritage Trail*—the Indigenous experience that has been provided by the *Goolarabooloo* and their Indigenous and non-Indigenous community members for over thirty years. During this nine-day walk, the *Goolarabooloo* represent Indigenous culture for an area whose Aboriginal groups are numerous and have traditionally been mixed (Glowczewski 1998). When walking, Trail participants cross through *Yawuru, Ngumbarl Djugun,* and *Jabirr Jabirr* country. Traditional linguistic differences and their respective territories are observed and *Goolarabooloo* culture is explained as a shared knowledge and shared understanding of country, unified by the practice of a single Aboriginal Lore and culture, known as the Northern Tradition (figure 1.3).

The way in which the *Goolarabooloo* practice their Indigeneity together with Trail participants is exemplary of how the preservation of

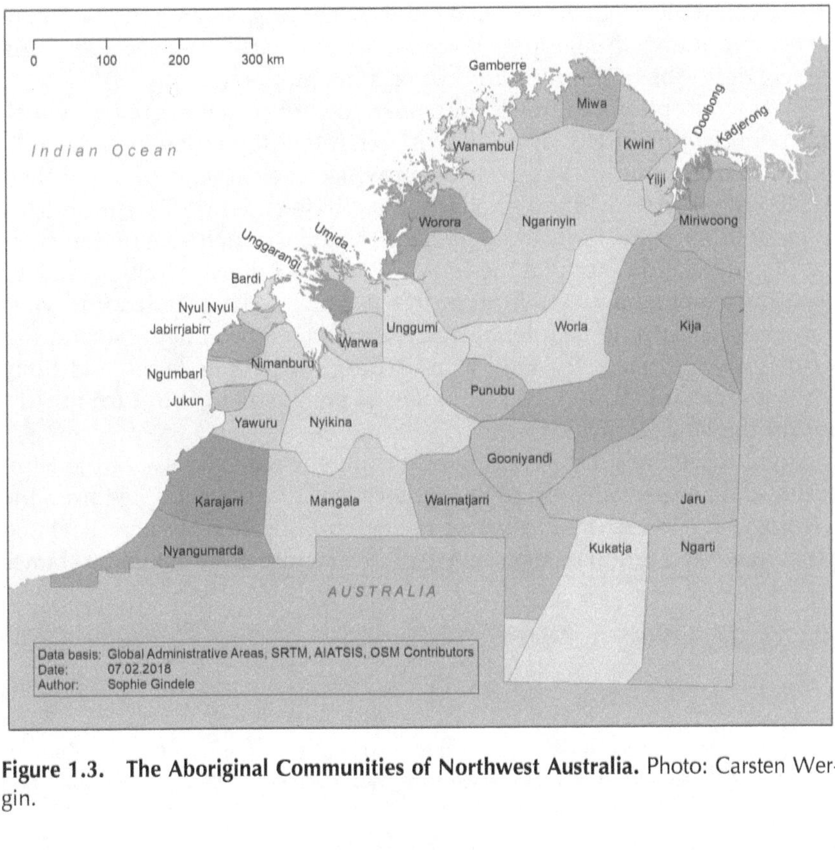

Figure 1.3. The Aboriginal Communities of Northwest Australia. Photo: Carsten Wergin.

Indigenous knowledge and country becomes a collaborative endeavor. Participation in the *Lurujarri Heritage Trail* offers the possibility of repositioning oneself beyond the "Indigenous versus non-Indigenous" divide. It offers a means for participants to become part of the hybrid cultural model that is *Goolarabooloo*.

The Walmadany/James Price Point industrial proposal challenged the value of this hybrid model to such a degree that a large group of people formed in opposition to it. Many of these people had become part of the *Goolarabooloo*. As will be seen below, they therefore wanted to protect not only the land but also themselves. The conflict became existential for them, because what the industrialization proposal put into question were values they had learnt to cherish through (participation in) the Trail and adopted as part of their modes of existence.

This book presents my ethnographic analysis of the diverse collaborations that stemmed from this Trail experience, and that prevented the LNG facility from literally *taking place*. Its introduction is followed by a discussion of key concepts such as the aforementioned notions of indigeneity, collaboration, heritage, resources, and values in the context of Settler-colonialism. I will also elaborate on the original notion that I generated out of my ethnographic material, which I term "transecology" and loosely define as transculturality beyond the human (chapter 3).[6]

These concepts form the overarching theoretical landscape in which this book is embedded. The following chapters then take as their individual starting points diverse ways in which the *Goolarabooloo* and their collaborators term themselves attached to "living country." Central to the collaborations is the *Lurujarri Heritage Trail* and the inclusive understanding of people and place that the *Goolarabooloo* describe as "living country" (Roe and Hoogland 1999). I found this notion particularly helpful when it came to imagining people, heritage, and the environment beyond dualisms. It emphasizes the land as nurturing for people, fauna and flora, while a *LORE*ful (in an Indigenous sense) engagement is in turn nurturing for the land itself.[7]

Along these lines, I will argue for a different world(view) of the Kimberley and beyond: a world of the "otherwise" that is uncoupled from the teleologically minded futurism of mainstream economics and a related belief in what might never be, like the LNG facility at Walmadany/James Price Point.

NOTES

1. "We are defined not by race, religion or culture, but by shared values of freedom, democracy, the rule of law and equality of *opportunity*—a "fair go." The

glue that holds us together is mutual respect—a deep recognition that each of us is entitled to the same respect, the same dignity, the same *opportunities*" (Malcolm Turnbull MP, Prime Minister of Australia 2017, *emphasis added*. Source: Multicultural Australia: United, Strong, Successful. Australia's multicultural statement, The Australian Government 2017. Accessible via the URL: https://www.homeaffairs.gov.au/mca/Statements/english-multicultural-statement.pdf (Accessed January 9, 2022).

2. Throughout this book, the term "Lore" has been adopted to make a clear distinction between Aboriginal/Traditional/Customary law (Lore) and Western/Modern/Settler-colonialist law (law).

3. For example, in the case of Walmadany/James Price Point, environmentalists would support Indigenous people even though the latter might offer sea turtle meat during meetings. Meanwhile, Indigenous custodians might disclose some knowledge about the land to environmental scientists that would otherwise be secret-sacred to them.

4. As of June 30, 2016, the Australian Bureau of Statistics estimated the Aboriginal and Torres Strait Islander population of Australia at 798,400 people, or 3.3 percent of the total Australian population. This estimate represented a 19 percent increase in the Aboriginal and Torres Strait Islander population estimate from that of 669,900 on 30 June 2011. Source: http://www.abs.gov.au/ausstats/abs@.nsf/mf/3238.0.55.001 (Accessed November 9, 2021). Meanwhile, in the 2016 Census, there were 16,222 people in Broome (Local Government Areas). 4,571 of these were recognized as Aboriginal and/or Torres Strait Islander people, making them 28.2 percent of the population. Source: http://quickstats.censusdata.abs.gov.au/census_services/getproduct/census/2016/quickstat/LGA50980n (Accessed November 9, 2021).

5. His name is given as Mr Roe throughout this book, in accordance with the Aboriginal Australian prohibition on speaking or otherwise using the names of the dead.

6. Philippe Descola and Gísli Pálsson (1996) first suggest the term "transecology" at the end of the introductory chapter of their famous coedited volume *Nature and Society*, with reference to Bakhtin's notion of "translinguistics," "to underline the notions of dwelling and embeddedness with respect to the human household, the social nature of the human *oikos*" (ibid., 19, emphasis in the original). Though there has been no further elaboration on their part, I content that my use of the term will at least partly link up with what they had in mind.

7. This book is thus also an ethnographic exploration of how debates about transcultural heritage and environmental preservation challenge mainstream notions of justice, the acceptance of difference and the acknowledgement of Indigenous rights in Settler-colonialist Australia.

Chapter 2

"Nowhere Else But Here"

On my first *Lurujarri Heritage Trail*, in 2012, I met a lecturer from the Royal Melbourne Institute of Technology (RMIT). It was his fifth Trail. He had a one-person tent and made use of it whenever possible, reading and relaxing after the longer walks. The third night on Trail, we sat at the campfire and I asked him about how he prepared his students. He responded that he never told them much in advance about what would be happening during the nine-day walk. "It's an experience that is difficult to put into words," he tried to explain.

He then added that usually, around the third or fourth night on Trail, one or another student comes up to him and exclaims in frustration, "I'm not learning anything." This is after three days of walking, following Indigenous guides through the land, accompanying them through mangroves to catch mud crabs or over the extensive dune system along the Indian Ocean coast, listening to them telling stories and pointing out middens and artifacts.

I asked how this was possible, and he explained that the way things are presented to the students on Trail is different from what they are used to. At this point, some students do not identify the experiences offered to them as learning until they are presented in an itinerary—until someone or something tells them when to switch into learning mode. This, however, was not how the *Goolarabooloo* would engage with Trail participants.

I found that the uncertainty the lecturer generated by not explaining much to his students in advance of the Trail was the most valuable methodological preposition he could offer them. The frustration with presumably "not learning anything" marks a decisive moment, as it sensitizes students to the world-making activities that they are engaging in already. Admitting to "not learning anything" means opening up to the possibility of learning differently.

After those first days on Trail, students and other participants let go. They no longer try to note down the take-home-message of the extraordinary university course they initially thought to attend. Rather, just as they stripped off layers of clothing in the heat, the third or fourth night is the moment when their insides strip bare, as well—this time not bare of clothes but bare of expectations, Western concepts, and other Modernist constraints. They are red-dirty from the *pindan* sand and soaked in sweat from the sun. They have learnt to seek the shade when sitting down with the *Goolarabooloo* to hear a story about the land and their community (figure 2.1). Slowly, routine makes way for the lived experience that everything is entangled—the people and the land. "This is very real," one student says. "This is living country," the *Goolarabooloo* explain.

This ethnographic moment from the *Lurujarri Heritage Trail* describes a key step toward the understanding of Australian Indigenous world(view)s and how they "endure" with both enormous creativity and enormous vulnerability. It requires a physical and mental bareness that is achieved through friction and consequent shock. The shock-encounter of the

Figure 2.1. Trail Participants Rest in the Shade during Story Time. Photo: Carsten Wergin.

students was triggered by the realization that they would not be learning in Western terms and therefore feared that they lacked the skills to learn anything. The shock-encounter for me was of a different kind and in a different setting.

Back in Germany, I had achieved positive results with my application for third-party funding to conduct this research project. But I was not granted the "ethical clearance" necessary to commence the project until certain issues had been resolved. The amendment process that followed was, it turned out, not a precondition for the fieldwork I was to undertake, but the beginning of it.

These two different cases of shock are points of entry into the methodological approach I took toward the field, because they both give evidence of how to productively engage with multiple and often unexpected challenges—what Elisabeth Povinelli (2016) might term "toxic" events—posed through lived experience, be it the experience of a participant during the Trail or of an anthropologist between jobs, in a financially precarious situation, with a desperate need for funding that was and at the same time was not granted. The question I pursue with regard to those shock situations is: How did I, and how do others, become part of collaborations?

The philosopher Jonathan Lear (2006) has argued that there is something more fundamental to the victory of one group over another than the question of who gets to write history. The "radical hope" that emerges when one's own culture and heritage is challenged to the point where it is subsumed by another is linked to a methodological question: will there still be appropriate concepts available to get the story told? Students who after three days of *Lurujarri Heritage Trail* experience ask why they are not learning anything are a case in point. What they have learnt no longer seems valid, but will they find the means to learn differently?

A successful Trail experience depends as much on those willing to share their story as on the ability of participants to literally "come to terms" with what their Indigenous guides offer them. By not attempting to explain what the students can expect from the Trail, the teacher makes way for a transecological experience, which does not add to familiar ways of understanding but challenges them instead.

I perceive my fieldwork experience in a similar way, as something from which I could not draw anything ethnographically if such ethnography were for the sake of representationism alone. Because of this, the methodological approach taken in this ethnography cannot but be self-reflexive (Marcus 1995). I therefore describe my access to the field and the material generated, as much as my involvement in the conflict, from an "emic" perspective—that is, as close as a white anthropologist of German descent can come to tapping into this perspective.

Conversely, there is no itinerary. No list of daily events that one can follow, no handout that sums up the most significant facts to be learnt. This book too is stimulated by "living country," where conventional learning experiences are discontinued. It is a challenge to that which is familiar, that which one has learnt—a challenge to conventional norms, values, and ethics. It might frustrate at times, as the first encounter with "living country" does Trail participants, who I hope will nevertheless stick with their (Indigenous) guide(s) rather than return to the comfort found in representationism.

University education is part of these conventions, and falls short because of them—or, as the Indigenous custodians explain to Trail participants from the top of a sand dune and with an overarching gesture, "You come from your university in Melbourne but this here is my university." The problem that this has turned into is how to "speak well" about "living country" and living heritage without falling back into conventional representationisms.

I borrow this expression, to "speak well," from Bruno Latour, who describes the act of diplomacy as "to speak well to someone about something that really matters to that person" (Latour 2013a, 46). The task for what Latour terms a diplomat—in his *Inquiry into Modes of Existence* he uses the example of an ethnologist addressing an informant—is to identify values and initiate negotiation processes based on how the reference chains that these values rely upon are filled with reality. Speaking well is thus a skill of articulation that makes utterances and different world(view)s comparable with a consciousness of the fact that reality can never be fully grasped in its representations alone. This capacity in a speaker leads to a diversity in which reality and plurality no longer stand in opposition to each other, but are part of the same experience. In my inquiry, this experience is the land, "living country" and living heritage: the *Lurujarri Heritage Trail*.

Meanwhile, my ethnographic experience began well before I walked the *Lurujarri Heritage Trail* for the first time back in 2012. As mentioned here, I had received on the one hand positive feedback on the third-party funding application I had submitted to conduct this project, but on the other was required to add significantly more detail in regard to my general methodology, and concrete permissions from local institutions, to obtain the necessary ethical clearance to commence my research. Working to meet these additional requirements turned out to be an eye-opener for me, as it demonstrated the entanglements of my own discipline of anthropology and its funding bodies in Modernist constraints.

The following section is structured around two stories that demonstrate how my frustration about the considerable extra work I had to engage in kick-started the project, which in retrospect made for a life-changing experience even before I entered the field. It is dedicated to the granting of ethical clearance in a double sense of the term: ethical clearance from the institutions that I was accountable to, and from Modernist terminologies that would limit my means of engagement with the world of the *Goolarabooloo*—my means to "speak well" about them.

2.1 ETHICAL CLEARANCE

In 2011, back in Germany, joy over the approval of my project proposal was only short-lived. Attached to it was a demand from the European Research Executive Agency (REA) that I thoroughly revise my ethical considerations. It appeared that I was given unlimited time to provide the necessary amendments.

I needed approximately four months, setting out from ethical guidelines provided by the university institutions that had agreed to host my project. I also consulted guidelines from other international research agencies and organizations I was a member of, such as the European Association of Social Anthropologists (EASA), the German Association of Social and Cultural Anthropology (GASCA), and the Australian Institute of Aboriginal and Torres Strait Islander Studies (AIATSIS).

Concrete requirements were not spelled out by the REA, but I assumed that one of them must be formal approval from the ethics committees of the two hosting institutions in Germany and Australia. To secure such approvals is commonplace for researchers in Australia. But it was a blind spot for me and, as it turned out, many of my colleagues in Germany, where approval for social and cultural anthropology projects is seldom in need of clearance from an ethics committee. As a result, no one could point me to the relevant ethics committee at my host institution in Germany. I eventually turned to the one based in the Faculty of Medicine, where ethical clearance is sought on a regular basis, for example for experiments with animals or human DNA.

Alongside my successful application to these ethics committees in Germany and Australia, I produced an additional document that accompanied my original research proposal, in which I guaranteed that all work would be conducted under strict ethical considerations. For example, I stated that names of informants and the personal data they provided would be "kept in close confidence" and "locked up in filing cabinets," inaccessible to anyone but me and my two "scientists in charge" (the official terminology of the funding body) at the host and return institutions.

Based on the information provided, I also produced an informed consent sheet in "lay language" including my contact details, which I later handed out to and had signed by my collaborators, together with an additional article that informed them of their unconditional option of "revocation of consent." The ethical clearance procedure therefore ensured that I was aware of relevant protocols, and provided my funding body with documented proof of my commitment to them.

What I became aware of in retrospect is that this process did not so much precede my fieldwork as lead directly into it. The reason was that obtaining ethical clearance also required that I provide letters of support from the relevant Indigenous groups I was to work with: first the *Goolarabooloo* community as custodians of the *Lurujarri Heritage Trail*, and second the *Yawuru* people as the Native Title holders for the Broome area. I had developed the project partly in collaboration with the *Goolarabooloo*, so the first letter of support was quickly obtained. The *Yawuru*, however, were only granted Native Title in May 2011, which was after I had submitted my original application to the funding body.

My attempts to secure a letter from them led me to the heart of a significant conflict between *Goolarabooloo* and *Yawuru* that is linked to the Native Title claim over Broome. The successful claim was originally submitted by the "Rubibi Claim Group," made up of *Yawuru*, *Goolarabooloo* and *Djugun* people. For what were termed strategic reasons, lawyers recommended reducing the number of applicants named to one official claimant, the *Yawuru* people (Burke 2011). The claim drew on knowledge that was preserved by Paddy Roe (OAM), *Nyigina* man and *Goolarabooloo*. After the successful determination, Mr Roe and his grandson remained on the *Yawuru* Board, because he was also recognized as *Yawuru*, but not as the representative of the *Goolarabooloo*. Ever since, *Goolarabooloo* and *Djugun* people have considered themselves underrepresented and/or ignored in decision-making processes within the *Nyamba Buru Yawuru Prescribed Body Corporate* (*Yawuru* PBC), the not-for-profit company owned by the native title holders, with "the responsibility to generate income from Yawuru's capital assets; land, community, social and cultural capital."[1]

My application had therefore put me, without realizing it, at the heart of an ongoing dispute. The need to fulfill my obligations to the funding body had catapulted me into my field site, but in a way that was still incomprehensible to me from where I was—at a computer desk in Germany.[2] Seeking approvals for my methodology pushed me into sociopolitical conflicts that not only shaped my field site but positioned me within it—with or without ethical clearance in hand.

At the time of my fieldwork, custodianship over the *Lurujarri Heritage Trail* was in the hands of Paddy Roe's grandson, Mr Roe. Due to the close relations my Australian "scientist in charge" had with the *Goolarabooloo*, it was a matter of hours to receive written approval from Mr Roe to conduct fieldwork with them. It was only in retrospect that I realized what it meant for Mr Roe to take the time to engage with my request. In 2011, the conflict over the industrialization proposal had reached another peak. Mr Roe was being pressured on numerous levels: industry and government questioned his authority as a cultural leader, and at the same time, he was being verbally abused and physically threatened on a regular basis in public by supporters of the industrialization proposal who identified him as the main threat to the economic opportunities that were thought likely to unfold with it.

Although I quickly received consent from Mr Roe, it took longer to receive the same from the *Yawuru* people. The then director of *Yawuru* PBC, Patrick Dodson, is known as "the father of reconciliation." Since 02 May 2016 he has been member of the Australian Senate as a Labor Senator for Western Australia. My project proposal was admittedly low on his agenda, for reasons that were understandable. During my stay in Broome, his commitments included not only that of the inaugural chairperson of *Yawuru* PBC but also leadership responsibilities in the Indigenous Expert Form, teaching obligations at the University of Notre Dame's Broome Campus, and the spearheading of a federal committee for the amendment of the Australian constitution to acknowledge Aboriginal Australians as "proper persons," known as the Council for Aboriginal Reconciliation.

I was lucky that Patrick Dodson had also recently become director of the Indigenous Policy and Dialogue Research Unit at the Social Policy Research Centre of the University of New South Wales (UNSW). The unit has since ceased to exist. However, at the time, it was the place where my project and I were hosted. This led the Dean of the Faculty of Social Sciences at UNSW to intervene in the process and ask Dodson to kindly take the time to provide his endorsement for my project to be implemented, which he did.

In retrospect, the problem is clear: I needed endorsement from both parties in the conflict, Mr Roe and Mr Dodson, to conduct research. But back in Germany, I was confined to the computer, and to emails sent back and forth with repetitive reminders and requests that remained unanswered. Meanwhile, I had to prepare my travels, organize accommodation for my wife, our seven-month-old daughter and me, obtain insurance, and fulfil visa requirements, including the obligatory chest x-ray for signs of tuberculosis at one of the doctors' offices approved by the Australian government.

I cannot name the day on which I succeeded with my funding application. I had submitted all of the necessary documents but never received an answer on whether those were deemed appropriate. Rather, the "go ahead" I expected to receive came in the form of never receiving a "no." The main proof of consent from both *Goolarabooloo* and *Yawuru* to write this book therefore remains the book itself, since numerous people from both groups, including Mr Roe and Mr

Dodson, were willing to meet with me and answer my questions, and contributed their time and energy to it.

Because of my work toward receiving ethical clearance, a thorough engagement with questions of translatability between Indigenous knowledge and so-called Western science became an important component of my research and methodological approach. This will become evident in my ethnographic chapters. The conflict over the industrialization of Walmadany/James Price Point can clearly be read as another failed attempt at translation between ontologically different world(view)s. But might it be possible to generate radical hope from such sites, as researchers engage in the multiple ontologies of human/non-human connectivity they provide for (Pahl-Wostl et al. 2013; Tàbara and Chabay 2013)?

To answer this question, I chose to rely on ethnographic practice, which as it stands, for me, describes the critical engagement with real-world problems under consideration of diverse ontologies, and through the lived experience of transecological diversity. Again, this definition emerged in the course of my fieldwork. I initially set out to experience "transcultural" diversity, since the notion of transecology that I develop in chapter 3 did not yet exist. Once I had arrived in Australia, I embedded my preliminary ideas and research results, including my experiences with the ethics application, into interdisciplinary discussion. I could draw on the expertise of anthropologists, archaeologists, scholars of cultural and postcolonial studies and environmental humanities, art and ethnohistorians, as well as geographers and, not least, policy researchers of the Social Policy Research Centre (SPRC), who worked with Indigenous groups on a regular basis (Cronin and Wergin 2012). In Broome, my fieldwork became more multisited through participant observation on different levels: within local networks of protesters and proponents of the project, in NGOs, at political events, in heritage institutions, and among individuals, (social) entrepreneurs, tourists, and travelers. The evaluation of preliminary findings was done in close collaboration with researchers and academics from affiliated institutions (Holm et al. 2013). For this, I used virtual discussions, internal review processes, and group presentation formats.

I conducted over 38 partly problem-centered, partly semi-structured interviews, sometimes for many hours and over multiple sessions, about a wide range of topics related to the conflict and with an equally diverse range of people. These might have been born in the Kimberley or have moved there, might have been employed by government or industry, worked within the Indigenous tourism and heritage sector, or come as visitors to the region and stayed. All these collaborators have provided vital information for this inquiry, which is why most interviews have been (at least partially) transcribed.

Meanwhile, I made a choice about whom to include in this book via direct quotations. My criteria were to accommodate gender balance and also a balanced point of view with regard to personal involvement with and connection to "living country," that is Walmadany/James Price Point. Not least because this is an inquiry into a protest movement that was generated by Australian civil society as a whole, Indigenous voices therefore equally feature with those of Broome locals, first-time Trail participants, non-Indigenous members of *Goolarabooloo*, tourism experts, environmentalists, and "opportunity dreamers."

Next to participant observation in multisited settings, the subjective knowledge of interview partners therefore became another primary source of data acquisition. As such knowledge can never be accessed in a complete sense (Honer 2000, 197–98), I compensated for the methodological challenge this posed using forms of "temporary membership." As a saxophonist, I became part of cultural practices. As a tourist, I participated in (Indigenous) tourism activities. As an academic, I presented at the University of Notre Dame Broome Campus and contributed opinion pieces to the local radio and newspapers. Last but not least, I produced a report for the *Goolarabooloo* that later became relevant in the decision-making process.

My goal from the outset was to address the intermingling ontologies of tourism, heritage preservation, sustainable development, Indigenous knowledge, science, and industrialization in an active engagement with local narrative. Having finally made it to the field, my first impulse was to eliminate the misunderstandings that had emerged in the course of my application process, in particular with regard to what turned out to be a frequent question: "Who do you work for?" People were suspicious as to whether I might be affiliated with government, industry, or other local stakeholders such as the Kimberley Land Council (KLC), all of which employ anthropologists to undertake their cultural and heritage assessments in relation to Native Title applications, resource exploration proposals, and other work to be conducted on Indigenous country.

My affiliation with two university institutions, one outside the Kimberley and one outside of Australia, and my third-party funding from the European Commission were helpful. The idea of a funding body from outside Australia supporting a project that was dealing with core issues relevant to Australian society generated additional interest and helped a conversation to begin. I could confirm that the project was my idea, and as such, independent from other institutions' interests or steering.

Such independence also provided me with considerable freedom with regard to whom and what I would or would not engage with. When drafting the proposal, I had found it convincing to propose my participation in the nine-day *Lurujarri Heritage Trail* as a valuable means of approaching the land around Walmadany/James Price Point and assessing its cultural and heritage values and potential as an Indigenous tourism product. The knowledge I had acquired because of this was helpful in the field, as it demonstrated to collaborators my dedication to people and topic.

The fact that I had brought my family with me was another significant door-opener. It was easier to become part of the community because of the numerous occasions on which fieldwork experience and daily life merged. For more than one and a half years, we integrated into Broome life, baked cakes for fundraisers, invited friends for dinner, joined bands to give concerts, and held play afternoons on our balcony for our daughter and her neighborhood friends.

On such occasions, a tourism promoter might happen to pick up his/her son from childcare the same time as me, which would leave enough space for a quick informal chat that could lead into arranging an interview. Likewise, I was both father and anthropologist when sitting at Broome's community center, Circle House, together with other parents whose children were playing in the adjacent room and outdoor area.

Twice a week, from 8 a.m. to 12 p.m. in the morning, I was the odd father sitting with six or more mothers, drinking coffee, chatting, working or reading. I sometimes brought my laptop, while others puzzled over accounting and taxation documents. Some might be so immersed in their conversations that they forgot my presence until one woman, whom I still remember from the day she decided to paint her toenails, reminded the others jokingly, "Careful, he is an anthropologist! He might write down everything we say."

I appreciated Circle House because it offered quite cheap childcare (compared to the average $100 AUS Dollar per day that childcare centers in Broome charged). It was also a place where I could rest from fieldwork, if need be. The conflict over the industrialization of Walmadany/James Price Point was not omnipresent there. In contrast, for my family and me it was a daily topic for obvious reasons. We scanned the local newspapers and noticeboards for related events and participated in them on a regular basis. We went to meetings with protesters at Broome Lotteries House. And while Environs Kimberley would unfold the results of their citizen science projects to a large audience, our daughter would play with other children, anxiously awaiting the promised sausage sizzle to follow.

We befriended *Goolarabooloo* families in our neighborhood, at whose places we frequently stopped on our way to the supermarket. At the same time, everything we did in town was complemented by visits to protest camps and participant observations of actions "on country." One of the places we frequently visited was the camp and information booth set up at Manari Road corner, maybe a twenty-minute drive out of town and as such a regular place to meet and chat, including with tourists who passed by (Wergin 2018c).

Manari Rd turns off Cape Leveque Rd, the access road to the Indigenous communities at Beagle Bay, Lombadina, and One Arm Point at the very tip of the Dampier Peninsula (figure 2.2). These offer Indigenous tourism experiences set in a pristine landscape, including mud-crabbing, spear-fishing, canoeing and storytelling around the campfire. Those who seek more individualized experiences take Manari Rd to find secluded camping spots and beautiful beaches along the Indian Ocean coast. Meanwhile, regular shuttle busses also frequent Manari Rd to take visitors to Willie Creek Pearl Farm.

As such, Manari Rd corner was an ideal spot for protesters in the Walmadany/James Price Point conflict to not only observe and generate attention but also form new alliances with locals, tourists and travelers. Large signs invited people to stop for a "yarn and a cuppa"—a talk and a cup of tea (figure 2.3). Protesters offered information on the conflict, images, and documents about the industrialization proposal and the history of the place in question and the opposition that had formed. They also

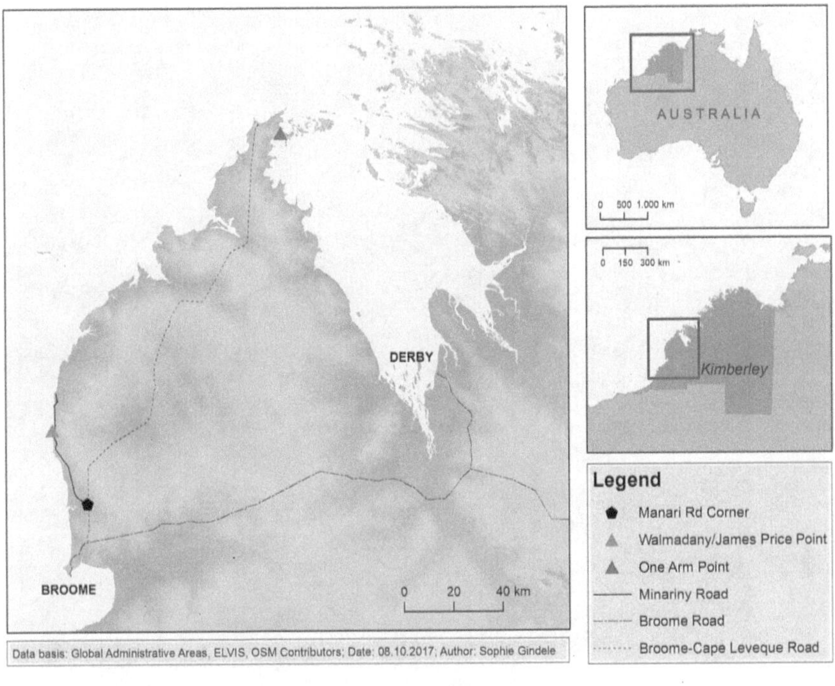

Figure 2.2. Map of the Dampier Peninsula, Also Indicating Manari Rd Corner. Photo: Carsten Wergin.

observed traffic to inform others in case of possible new attempts to bring heavy machinery on country.

I would speak with protesters at Manari Rd Corner about recent events. My intention was to get a sense of the energy and commitment it took to keep up the protest camp, not least because of the heat, mosquitoes, and continued conflicts with police authorities. The activities of the private security company Hostile Environment Services that had been hired by Woodside to protect its industrial compound were another issue.

The security company had been the target of some mockery, not least because of its name. Though heat, mosquitoes, and poisonous animals might make for a hostile environment, protesters would turn this name around, referring to the employees of the company simply as "The Hostiles." The latter would drive up in teams and white air-conditioned four-wheel drive, with sunglasses and baseball caps pulled low over their faces. One of them would carry a camera to randomly, and without prior "informed consent," videotape people at the protest camp, which also included my family and me.

Their outfits and expensive equipment protected them well from the so-called "hostile environment." The sun was strong, and the red dirt could reach every corner of your body. However, my family and I found

Figure 2.3. "Meet and Greet" Protest Camp at Manari Rd Corner. Photo: Carsten Wergin.

the camp at Manari Rd not hostile but welcoming[3]—even more so since the protesters who kept watch were usually retirees with the necessary time at hand, more interested in a cup of tea than violent upheaval. Quite a few of them were women who had lived in Broome for the last twenty to thirty years. They ticked some boxes in terms of stereotypes: old hippies and artists in wide-fitting, colorful clothes. As stated here, Broome is known for its hippie heritage and related lush, artistic lifestyle that continues to attract tourists and travelers to visit.

In sum, my fieldwork can be described as a puzzle out of which I picked up one piece at a time, without knowing how its outline looked so as to determine how it fitted with others. I knew that they did: *Goolarabooloo, Yawuru*, Circle House, Manari Rd, hippies, "Hostiles," the Old Hostel on Dora St . . . During fieldwork, these pieces did not produce a complex whole but continued to attach to other pieces. They did not make a system, but became knots from which I could direct my attention to various domains.

This concept of knots proved useful in grasping my fieldsite, because it also offered a means of describing how the protest itself was organized. In fact, knots rather than the systems-approach were what made the protest so effective and at the same time so confusing for anyone outside its networks. Patrick Dodson, for example, once asked a close associate who to talk to if he wanted to meet the leader of the No Gas campaign. The answer, "There is no leader," left him in disbelief.[4] How could the construction of a $45 billion AUS Dollar LNG facility that was advocated for by powerful entities such as the West Australian government and Australia's largest independent oil and gas company be delayed for over five years by a group without identifiable leadership?

It was a question that became central to my fieldwork, the methodologies I employed and the theories I generated. Martin Pritchard, director of the local NGO Environs Kimberley (EK) that played a crucial role in the successful opposition movement, described this group as follows:

> *Martin Pritchard:* It was a real grassroots effort in that people would meet to talk about strategy and media, other people meet to talk about logistics and non-violent direct action. There'd be meetings where people came together around the kitchen tables, or the dinner tables, to share what was going on. Surveillance was another strand. People would (. . .) communicate really closely on every different issue imaginable, every single bit of information that came through, and work out where things were heading. It's quite organic, and quite complex (. . .) basically like a web. That web would reach all the way out, and then there'd be people in the middle trying to gather all the information and put ideas together . . . . And put it into action . . . . Then the action spread out through the web again.

Carsten Wergin: It never seemed that there were people who would actually give orders. . .

MP: No. It was all about ideas . . . . And people working through the ideas, as to whether they were good or bad, and what the risks were . . . . What the opportunities were . . . . And bouncing them around . . . . Coming to an agreement, spreading that buck all through the web, getting feedback and then, "*fang*" it happened. (Transcript 130513_1_MP)

The hypothesis I began to explore was that the strength that emerged from these organic collaborations found its expression in the aforementioned knots. Thus, I began to search for them, and to be present in them whenever possible.

One of these knots forms when Trail participants leave Broome and walk along Cable Beach until they reach their first overnight stop, *Ngunungurrukun* (Millibinyarri/Coconut Wells), 15 kilometers north of the town (figure 2.4). On Cable Beach, trailers pass tourists who sit in the shade of rental umbrellas, while further on, four-wheel drive owners drive close to the Indian Ocean, get out their deck chairs, and watch the sunset with the privilege of an ice-cold beer out of their 12V refrigerators at hand. In such instances, the Trail participants, who cross into the imaginary "last frontier" on foot, challenge tourists and 4WD owners: what begins where

Figure 2.4. Trail Participants Pass Tourists on Cable Beach. Photo: Carsten Wergin.

the township of Broome ends, and certainly where the 4WDs on the beach cannot go further, is assumed wilderness. In crossing this frontier, Trail participants loosen their knot with the Settler-colonialist world, while the one with Indigenous Australia begins to tighten.

On the fourth day of the Trail there is a further string attached to this knot, when one of the guides exclaims, "There . . . the old fella" and points to a skull partly uncovered by the wind. The *Goolarabooloo* have been on the lookout for the skull. Each year, it is a highlight for Trail participants. With awkward looks on their faces, some ask, "Are we not disturbing the dead?" But curiosity prevails. Trailers circle around and take pictures in the early morning sun, which rises in the east to plunge into the ocean at the end of the day. After they have a look and before they leave, the *Goolarabooloo* cover the skull again—"proper way," as they say, for the next wind to expose the "old fella" again.

Walking on, Trailers hear more stories of a "living country" that strengthen the knot between the *Goolarabooloo* and their visitors. Slowly one grasps the meaning of why this land is sacred—a living heritage. This feeling is enhanced with the arrival at Walmadany/James Price Point, where in 2012 an eternal fire burnt (figure 2.5). People erected the very first of its kind in 1972 in front of Old Parliament House in Canberra, while their outpost, *Walmadany Tent Embassy*, was comparably young.

Figure 2.5. The Eternal Fire at Walmadany/James Price Point. Photo: Carsten Wergin.

The sign reads "14 September 2011." A No Gas corrugated iron sheet stands close by, ready to encapsulate the place in case of strong winds that might put its fire out.

Evidently, Trail participants have by now entered a "living country" where human and more-than-human actors care about an "old fella" just as much as the fire. Such knots are indicators of people and place coming together in an extraordinary way, in a "nowhere-else-but-here"—so much so that the WA Museum in Perth dedicated a section of its new Western Australian Museum to the Walmadany/James Price Point conflict.[5]

When my family and I returned to Broome in 2017, a community working group was set up as a contact for museum authorities to assist in the design of this exhibition space. I was asked to attend a couple of their meetings. Again, there was no leader, but only a collaboration of people who exchanged ideas about how to educate about the protest and tell its story.

It was the possibility of participating in such meetings as well as the feedback they generated from the community that helped me to verify my analysis. They also demonstrated the ways in which I had become part of collaborations myself. In their response to questions from the museum, the group defined the main characteristics of the protest as:

(a) serious campaign planning and a strategy that was constantly reviewed, refined and perfected, but without a conventional "command and control" structure,
(b) different parts of society coming together in a "tapestry of talents"—"Old Broome" families getting together with newcomers, Aboriginal custodians, traders, lawyers, nurses, doctors, checkout operators, teachers, business owners, bankers, academics, environmental activists, politicians, NGOs, charter boat operators, musicians, artists, music stars, actors, journalists, and
(c) a new language of resistance.[6]

Here, I have discussed features (a) and (b), and shown how both—collaborations and knots rather than hierarchies—have been fundamental to the protest movement and my methodological approach. In what follows, I turn to (c), "a new language of resistance," which was the language of love and harmony, which during the "Mother's Day Protest" of 2012 involved mothers and their children, roses for police authorities, and posters with hearts to express a love for Broome and the particular values that are already in place there. The following subsection addresses these particular kinds of "affirmative action," while also adding more detail about how my family and I became part of what Broome's iconic local band the Pigram Brothers describes in a song as a, "nowhere else but here."

2.2 COMING TOGETHER

Once I was in the field, the ethical commitment that I had pledged in the amendments to my project application kept reverberating through one particular question: what are you giving back to the community? I was eager to engage in opportunities to not only take in information, stories and data but also participate and contribute my own expertise and skills. One means to give back was provided by my saxophone, with which I joined local musicians for fundraisers and concerts, as I had done during previous fieldwork (Wergin 2010b). Other means included giving interviews with the local newspaper and radio stations, or presentations at Notre Dame University, Broome Campus. On such occasions I found myself wondering, how much of my opinion I could give without compromising my integrity as a researcher? How big a part of the collaborations could I become? How much "giving" was also good for my family—for our daughter, who entered childcare and formed her first friendships with people she was to say goodbye to at a foreseeable point in time? These are ethical questions that lead further than any research institution can give clearance for. They are attached to lived experience, immanence, and the everyday of an anthropologist.

Along these lines, the requirements for ethical clearance and my initial attempts to work through these both before and in the field paved the way toward another central concern in my inquiry: the question of values. Still in Germany, I planned to assess the cultural and heritage values of the area and its value as a tourism product from the perspective of the Indigenous group that organizes the *Lurujarri Heritage Trail* each year: the *Goolarabooloo*. Upon arrival in the field, however, I was confronted with a long history of conflict that culminated in the threat of the Trail's destruction in the form of having an LNG facility constructed on top of it.

Given all I knew, I was astonished. How could an LNG facility be built on top of my designated field site given the documented heritage values it contained? My research focus thus shifted from an assessment of the values attributed to the region by the *Goolarabooloo*, Trail participants and tourism entrepreneurs toward the Trail as a site of conflicting regimes of value, of "Songlines vs. Pipelines" (Wergin and Muecke 2012).

I became cautious about not turning this into yet another analysis of a dualism of the "evil" industry versus the "good" Indigenous people. In order to avoid this epistemological trap, I set out from the land as that which both people and industry, as well as all other diverging regimes of value, share an interest in. Following on from this, I began to explore the impact of Indigenous voices from local politics all the way up to supranational decision-making processes. Within the social fabric of Broome there was a rather clear understanding of who to ask for permission. This awareness grew stronger depending on how long people had lived in the community. People who had moved to Broome within the last twenty to thirty years might term themselves "locals" but did not dare to consider themselves "local local"; the latter expression was only used for an exclusive part of the local population referred to as the "Broome families."

These families joined the public protest relatively late, after "locals" had already formed the No Gas campaign. The first time I saw both groups work in collaboration was at a community event organized by the Broome families on the back lawn of the Our Lady Queen of Peace Cathedral in the heart of Broome's Old Town. The Pigram Brothers, who are famous across Australia for their particular musical style that reflects Broome and its surrounding country so well, played their iconic song "Nowhere else but here" (figure 2.6). Members of the Broome families took to the microphone and gave personal accounts about their family relationship to the town and what they held dear about it. Most participants in the audience had come with banners that showed hearts for Broome, or thumbs up for turtles. I noticed that everybody avoided mentioning the LNG proposal. It was not to become a gathering in opposition to it. It was a gathering to celebrate what was already in place.

A similar event had been staged a couple of years before at Cable Beach. The Pigram Brothers were again involved. They gave a free concert

Figure 2.6. We Love Broome Protest in the Church Yard. Photo: Carsten Wergin.

that they performed on board a catamaran called *Karma IV*, which usually takes tourists to secluded places along the coast. Ebony L., one of the initiators of the No Gas campaign, provides a vivid account of the event and its peaceful quality in this interview excerpt:

> A few days beforehand, we had a meeting out at the Crocodile Park. [Before this], Chris Maher and I delivered a petition to Howard R. [then Broome Shire President]. Howard said, "What's this about, this *event* on Cable Beach?" Chris and I looked at each other and said, "What event?" Howard says, "Just a minute." And marches off to an office. He comes back and he's got this printed page off Facebook, which describes this "event" on Cable Beach. Chris said to him, "Oh, Howard, that's just Facebook. Whenever someone has a party—at Facebook they call it an 'event.' It's not a proper event, Howard. It's a Facebook thing, that word 'event.' They're just going to be playing there." And Howard says, "Are they going to be playing on the beach?" "No," says Chris, "it's just going to be a bit of a picnic" (. . .) We wandered off . . . We could see he wasn't satisfied (. . .)
>
> We get out to the Crocodile Park . . . . And Jane Lawton offered to organize a bunch of people to deliver these flyers . . . . And the place was *plastered* with flyers (. . .) We put them under windscreen wipers, they got handed out at shopping centres... And it doesn't take Broome people much to get them going to a *free* concert. And if the Pigram Brothers are playing (. . .) So (. . .) police are scratching their heads because Howard has told them that there's a bit of a picnic . . . And we've got no security, we've got no extra wristbands, we've not organized any parking. We've said to people, "This is a picnic kind of thing" and "Bring your banners and your No Gas t-shirts." And people came with their dogs—"No Gas" written on it—and hats and their No Gas t-shirts and some had little banners. It was amazing. People just came from nowhere. I never told a lie. People came for a picnic and the Pigrams did play and they didn't play on the beach. They played on the boat in the water. We cleaned up the beach afterwards. There was no disorderly conduct, there was no assaults, there was no police arrests, there was no security issues and there was no rubbish (. . .) And there was seven to ten thousand people. I asked Alan Pigram and he told me ten. I said, "Oh everyone else is saying seven." And he said, "I've sang enough in front of big crowds. I know what ten thousand [looks like]." So, how many communities could you organize, one, to do that in three or four days, and two, not have a drama, and not have a mess to clean up afterwards? That's what I mean by peaceful. (Transcript 130326_2_NW)

The conflict evolved and more people grew aware of the wider consequences of large-scale industrialization close to Broome. Eventually, it became a full-on issue in local politics when a group formed and initiated a campaign for the Shire elections. The Pigram Brothers were again present, as candidates Chris Maher (aforementioned by Ebony L.), Anne

Poelina, Kandy Curran, and Mark Manado adopted the title of their song, "Nowhere else but here," as their campaign slogan.

By now, the song title-turned-political slogan had fully caught my attention. The different regimes of value, the methods for Indigenous representation in decision-making processes, the ways in which people demonstrated not against the LNG facility but for Broome and what was in place, met in this simple formula and the further lyrics to the song:

> Now I'm dreaming by this campfire, gleaming in this dear old land of AUS / Flippin' idly through the pages of the tales of the never was / Losing interest in a future what with tomorrow so unclear / I guess maybe I'll never really need to be nowhere else but here. (The Pigram Brothers 1997)

My focus therefore gradually shifted again, from regimes of value that played out against each other to the particular qualities that the conflicting parties saw in the region: the resources industry, the Indigenous custodians, the environmentalists, the tourists and travelers, and the Western Australian government, whose premier Colin Barnett was determined to build the LNG facility "nowhere else but" at Walmadany/James Price Point.

The additional work I had to engage in to receive funding for my project—notably the ethical clearance procedures I needed to tackle—had "kick-started" my engagement with local Indigenous and non-Indigenous authorities. It had raised my awareness of the complex history of Broome and its people, who are at once divided and collaborative through their shared commitment to their town. Now, on site, the protest brought the qualities that people saw in Broome to the fore in extremely creative, peaceful and collaborative ways. On the other hand, my fieldwork also saw me engage in equally creative but potentially less peaceful means of protest. These played out "on country," at Walmadany protest camp, Manari Rd Corner and on the access road to the Woodside compound, and culminated in a clash between protesters and special police forces on what became known as "Black Tuesday" (Muir 2012). Martin Pritchard of Environs Kimberley remembers the impact of this day as follows:

> The camp was set up at Walmadany in April 2011, I think it was. Then, a couple of convoys were stopped on Manari Rd. There was one particular convoy that was stopped (. . .) and that was huge in terms of people mobilizing. A decision was made to, rather than trying to stop them between Quondong and Waterbank, take it back to them . . . Go back to the corner of Manari Rd and stop them there, closer to town. 2WDs can make it there, people can camp there (. . .) That mobilization happened before, but I think Black Tuesday brought it to national attention. There were media crews flying in from Perth and Sydney. [Black Tuesday] might have been a month of

> build-up . . . . It was pretty well planned and quite strategic in its aims (. . .) Meetings would be called and there was one meeting just via the No Gas phone, "We're concerned . . . . We've seen lots of cops in town . . . . We need to have a meeting in our little corner (Manari Rd) tomorrow at four o'clock." That was on a Saturday and the meeting might have been on a Sunday . . . . And you'd have eighty to a hundred people turning up for that meeting, prepared to mobilize physically (. . .) Obviously that's when the media and everyone caught on to it in a bigger way. (Transcript 130513_1_MP)

What first seemed a rather obvious dualism of "Songlines vs. Pipelines" became a complex web of various regimes of value, of locals and tourists, industry and politics, environmentalists and Indigenous custodians, that crisscrossed in multiple ways. From there, it went on to become a fight for media attention and public support, finding its peak on "Black Tuesday" when special police forces that had been flown in from Perth escorted trucks with machinery for the compound and were confronted by more than a hundred protesters on Manari Road.

Many of the local community members who had joined the protest were traumatized in the aftermath of this partially violent confrontation with police authorities. In order not to lose myself in a widespread anxiety that such an event could repeat itself, I turned my attention back to what initially had brought me to Broome, the *Lurujarri Heritage Trail*. The Trail once again became my means to stay "earthbound" (Latour 2017). I moved away from documenting knots, moments of crisis, and protest actions, and toward learning about the qualities and values ascribed to the land and transmitted in this experience of its living heritage. In other words, whenever I was spinning too far away into the web, I returned to the Trail. The Trail preserved both "living country" and myself from the discursive powers of whatever interest group I was tracing, so that I could pull myself out before I became subsumed into them.

This was at a point in time when Broome had become a home for my family and me. We had established a personal connection to the place and an equal reason to describe it as our "nowhere else but here," while government and industry argued for building an LNG facility nowhere else but here, even though Walmadany/James Price Point was not necessarily the best possible place for it.

The decision to build at Walmadany/James Price Point was not taken solely based on best practice assessments, but partly as a result of power politics. It showed that Broome had remained a Settler-colonialist town, not least since its Chamber of Commerce hosted the 2012 Kimberley Economic Forum "Owning the Future," with no Indigenous contributors except for Jimmy Edgar, a *Yawuru* and *Karajarri* Elder, who gave the Welcome to Country.[7]

This is how the final piece in my methodological puzzle fell into place, which was the engagement with Australian Settler-colonialism in its contemporary form. I came to ask how Indigenous people managed to safeguard their world(view)s against the powerful discourses of progress, multiculturalism, and development. How did they endure? The answer I found during my fieldwork was: in collaboration.

2.3 BECOMING PART OF COLLABORATIONS

My definition of collaboration aligns with what Li (2000, 151) suggests as a hybrid understanding of Indigeneity, "a positioning that draws upon historically sedimented practices, landscapes and repertoires of meaning, and emerges through patterns of engagement and struggle." I term this definition hybrid because it leads beyond the conventional use of Indigeneity as a draw card to "justify and defend a desired status quo" (Weaver 2016, 31; see also Wergin 2012). In other words, those who become part of collaborations in defence of "living country" commit themselves to preserve an alternative world(view) outside of the neoliberal mainstream. The motivations for such commitments are of central importance to my research project and thus became the focus of my attention.

My first visit to Broome was in October 2011, when I arrived together with Stephen Muecke, who the European Commission had termed my Australian "scientist in charge." We stayed at Notre Dame University for the first couple of nights. It was dark and the town was difficult to navigate. Broome is set out in squares like a typical colonialist town, but because it is adjacent to Roebuck Bay, the streets partly follow its meandering coastline. Since Stephen's last visit, Broome had fully metamorphosed into a tourist town, with campsites, cafés, roundabouts, palm trees, parks, and two shopping complexes that host two large supermarket chains (Coles and Woolworths), the department store Target, the electronics chain JP Morgan, and other Australian retailers.

Our plane landed just after 6 p.m. and I remember a hot, dark and unfamiliar town with what seemed to be no one in the streets. When we arrived at the university reception, no one was there, either. Adjacent to an open communal kitchen space we found a small apartment, and upstairs a dormitory, both open and reserved for us. Both facilities had air-conditioning, and the apartment was equipped with a kitchen and en suite bathroom and shower.

Travel had been exhausting. Direct flights from Sydney were already scarce in October, as the tourist season had almost ended. We had flown five hours from Sydney to Perth, followed by a two-hour stopover, and a final two-and-a-half-hours flight to Broome. After freshening up, we went to the Japanese restaurant *Azuki* for dinner and met with Howard Pedersen, coauthor of *Jandamarra and the Bunuba Resistance* (2000). The book tells the story of the Kimberley Aboriginal warrior Jandamarra, who hid for over three years in Windjana Gorge and Tunnel Creek near Fitzroy Crossing, from where he fought the pastoralists and colonial troops until he was killed on 01 April 1897.[8]

The restaurant *Azuki Japanese Fusion* is located on Napier Terrace, opposite to Male Oval at the entrance to Broome's Chinatown. Next to it is a car rental and another Asian restaurant that seemed to be closed when we got there. Howard was fond of the Japanese restaurant, but the prices on the menu were a downer for me. I opted for some snack-like dishes and wine, while the other two had sushi for starters followed by soup and sweet-and-sour beef.

The food was tasty and the place nicely decorated. It was crowded. There was faded greenness from the grass on the oval and some dim, yellowish streetlights. The obscure darkness was enhanced by my muddled brain, which reacted to the hot and humid climate with memories of previous fieldwork experiences at the other end of the Indian Ocean, on the French Overseas Department La Réunion, Mauritius and Rodrigues (Wergin 2010b, 2015, 2017a).

At this stage, I did not comprehend how different Broome was from where I had previously experienced such weather. What I quickly grasped, however, was how different it was from the rest of Australia, because of its long history of a living Asian cultural presence, as Frans Hoogland explains:

> *Frans Hoogland:* For Broome people, [the town] was connected (. . .) much more to Asia than to anywhere else in the world (. . .) There were people from (. . .) Malaysia, from Kupang, from Japan. So, it had its eyes on Asia already. It was fairly connected. And I remember all these young musicians who are now kind of famous, like the Pigrams . . . . Because there was no fences and there was no television, so everybody was having parties. Coming together and having a good time together . . . . And Jimmy Chi writing his play "Bran Nue Dae" in those days . . . . of the hippies (. . .) But we knew that the whole world would come (. . .) [laughter] It's got something, this place. It's cosmopolitan. It has got everybody here walking around from different nationalities, on a pretty good level. It is not paradise . . . . There's still human drama, but [it] could be a lot worse (. . .) And then the world came here . . . . First the [sealed] roads so all the cars could come drive all the way (. . .) Hippies first, and the travellers first. But, there was good people in those days . . . . No much damage in the country and a lot of people were really well. Respectful for Aboriginality. There wasn't that racism (. . .)
>
> *Carsten Wergin:* People are saying that's still this way today (. . .) It's quite unique for Australia.

> *FH:* Yes, we are a long way away from Australia basically. (Transcript 120807_4_FH)

Similarly, that evening at *Azuki* I was in Australia but felt a long way away from how I had got to know it thus far, in our first weeks in Sydney. The following day, Stephen and I went to Cable Beach for coffee and saw the Indian Ocean for the first time. We sat in *Zanders* on what I thought was a cliff but was actually the remains of a dune formation on which the café, a playground, and a large lawn with seating, paths, and fences had been built. The dune formation had been bulldozed to further accommodate a large car park, another playground, and a bar next to the lifeguards' house. I could identify smaller trees and bushes but no particular vegetation. Everything seemed vast and bare, like the parking lots near the shopping centers in town.

These impressions were accompanied by a conversation I had with Stephen about center/periphery distinctions. In one of his books, the sleeve note speaks about a dinner experience he had with Jacques Derrida, where he was asked by the philosopher, "Why the center?" After our dazed arrival in the dark and disoriented search for Notre Dame University, we came to the conclusion that in Broome this question might translate to, "Where the center?"

The answer was given to me some eight months later. Upon my return from the *Lurujarri Heritage Trail*, after nine days of walking through "living country," experiencing a living heritage, I became convinced that the center was not the town but the ecology in which it is embedded. I had been attracted to Broome because I wanted to see another side of the Indian Ocean that I imagined as being at the "opposite end" of where I had spent most of my previous fieldwork years. I was looking for change and expected Broome to be different. Such difference can be in a feeling. It can be a place. It can be a sound. It can be an encounter with another person. It can be a thought. In Broome, it is the experience of "living country."

Rather than being defined in opposition to something (like the opposite end of the Indian Ocean), "living country" defines itself and finds expression in multifaceted collaborations between the human and the more-than-human. It has inherent qualities that attract attention and are valuable in themselves. Camping sites and sacred places, protest actions, heritage sites, stories of the *Bugarrigarra* knit together a place to which the resource boom—the very foundation of Settler-colonialist Australia—has been a concern not as an opportunity but as a considerable threat. "Living country," however, had thus far prevailed.

During my fieldwork, everyone recognized an inexplicable assurance coming from the *Goolarabooloo* that this would also be the case with regard to the LNG proposal for Walmadany/James Price Point. This assurance, which stemmed from "living country," impacted the protest actions as much as on my own work as an anthropologist. I conclude this chapter with brief descriptions of further knots that became central sites of personal engagement and collaboration with the local community for my family and me.

2.4 ON DORA STREET

The Old Hostel in Paddy Roe's block in Broome's Dora St is the starting point for Trail participants. My family and I lived across the street from the block, in a two-bedroom apartment unit owned by the Aboriginal corporation *Mamabulanjin*. It was not easy to find the first-floor apartment. Most places of its size were charged at around $600 AUS Dollar per week rent. Prices in Broome were up because of the looming mining boom that was about to hit the town.

We were able to secure the place thanks to Stephen Muecke's contacts, and the support of local stakeholders such as the then CEO of *Goolarri Media*, Kevin Fong. It was owned by *Mamabulanjin* and usually accommodated their staff. But if there was availability, they rented it out to other companies. In our case, the apartment was rented out to *Goolarri Media* and subleased to us.

With regard to the empowerment of Broome's Indigenous population, Paddy Roe and his block on Dora St play a key role. The block is located just outside what was known in the early days as the Common Gate—the town's boundary, which Aboriginal people were not allowed to cross after sunset unless they had a special permit—for example, to work on freshly harvested pearl shells. The grounds became the site for regular *Corroborees* and other meetings, animated by Roe and the famous *Nyigina* artist and Songman Joseph "Butcher Joe" Nangan (Jahn and Drummond n.d.).

From the 1980s onwards, Roe also provided parts of this land for different organizations to use to establish themselves. The most important of these, which exists until today and has its head offices on Dora St, is *Mamabulanjin*. Its CEO, Paul N., retells how it came about in these interview excerpts:

> *Paul N.:* "Mamabulanjin" means "Meeting Place" (. . .) So that already conjures up in anyone's thinking, I suppose, that there was a need for Aboriginal people to be able to meet (. . .) It was incorporated by a group of like-minded people that thought that Aboriginal people needed representation (. . .) A lot of the Aboriginal people were herded in by the missionaries . . . just over a hundred years ago and herded into missions: Beagle Bay Mission, Lombadina Mission,

One Arm Point Mission, and so on. Even my mother was a part of the Beagle Bay Mission back in the late 1930s. Those people were dispossessed. They were taken from where they camped and lived and where their fathers or mothers were born, and herded into the missions. *Mamabulanjin's* role was to try and get people back to their historical or traditional homelands. Paul Keating was the Prime Minister at the time that introduced the Aboriginal Community Development Policy [ACDP] (. . .) [from] which CDP came about. And the first lot of ACDP funding was the Homelands Movement—it was labelled then (. . .) That was run in conjunction with CDEP (. . .) which was the employment and training component. And so ACDP was the capital funds that purchased generators, bores, tanks, pumps to get the people back out onto country . . .

Carsten Wergin: The infrastructure . . .

PN: That's right. And so on . . . developed tourism and other projects, you know. Paul Keating at the time was the one that was trying to induce the government to say, "Sorry" (. . .) It was the mid to early 1980s where all that was happening. And I suppose the people that decided to create *Mamabulanjin* was riding on the back of CDEP, and what Paul Keating was saying . . . a precursor to the Mabo decision, which came later on, four or five years later (. . .)

CW: Who were the people who set it all up?

PN: There's our corporation certificate [points to the wall in his office] which says 2nd of August, 1985. Paddy Roe opened this building back in December 1987, I think. So a couple of years after incorporation (...) Kees Pley, Peter Yu, Stephen Pigram were the first ones that applied for the funding to start it all off. And, actually, all members of the inaugural corporation paid fees (. . .) That was a commitment by the people (. . .) They were all financial members, which is quite a big difference to nowadays, because you don't generally have that happening (. . .) All these bricks are all handmade by local Aboriginal people! All this brickwork is red, rectangular, pindan brick. It was never engineered. It was never endorsed. It was never approved. But we got it under the radar, and they bought a brick-making machine (. . .) I think after they built this building, they moved the brick-laying [*sic*] machine up to Kennedy Hill, and that community hall is built out of the same brick (. . .) I'm pretty sure that's how the story goes.

And so from humble beginnings and, I suppose, a very labor-intensive program—you know, it was an Employment CDP program, one of the most viable projects around, back in the 1980s (. . .) So, *Mamabulanjin* was started basically to service individuals and corporations, because if you wanted to get ACDP funding and CDEP funding, and you wanted to move back to your historical or traditional homelands, you had to get incorporated. And so it was a *mad rush* back in those days to get incorporated under the federal legislation . . .

CW: So, would *Mamabulanjin* help?

PN: That's right. They would help them with the requirements under that legislation for the incorporation . . . . If they got funded they had to do returns. They had to maintain a register. They had to have regular meetings, and *Mamabulanjin* did it all for them, basically.

CW: What were the corporations that started at that time? Are they still around today?

PN: They're all around. The first one was *Nyul Nyul*, so that's beside Beagle Bay. That all started through here. Then there was *Nirrumbuk*, (. . .) *Nyul Nyul*, Pender . . .

CW: Pender Bay?

PN: "Pender Gardens" it was called back then. Pender Bay is actually a garden lease—a nursery lease, and that's how the family group got their foot in the door. And that was one of the first places that we could fund, because they had land tenure through that lease . . . . If you know Pender Bay's history, that's where they had the police first, it was a police station. What was it . . . a hundred years ago (. . .) It was being developed as a port. But Beagle Bay ended up becoming the port (. . .) Luggers used to pull up there . . . . Not a port, per se, because the tide's too big (. . .) You had to have flat-bottom boats that could beach themselves, and most of the luggers could do that. Well, actually no, luggers had a rounded hull. But they'd always pull up the side creeks because of the natural "V" shape of the creeks. But Pender Bay had a police station . . . . My son's great-grandparents, I think, were related to the policeman. That's the Masuda family (. . .) They're connected to the Bells that were the first policemen out at Pender . . . quite a long time ago. (Transcript 130408_1_NG)

This brief recounting by Paul N. of the history of *Mamabulanjin* ties various knots together. First, it highlights the importance of Paddy Roe as an Indigenous Elder with a significant vision, who strongly supported initiatives for the self-empowerment of the Aboriginal population. Second, the fact that the very offices of *Mamabulanjin* were built by Indigenous people proves false any assumption that Aboriginal people will not commit themselves, as long as they find themselves represented in what they commit themselves to. And finally, the establishment of Pender Gardens/ Bay as one of the first projects supported by *Mamabulanjin* ties Aboriginal history to that of Japanese pearlers such as the famous Masuda family, whose patriarch came to Broome in 1955 and whose life story is part of the pearling legends around hard-hat divers of the time.

Dora St is named after a horse that in the early 20th century had three consecutive wins at the Broome Cup—another historic attraction for domestic tourists, who use it as a reason to travel to Broome for the weekend. Meanwhile, like the Courthouse Markets or the Staircase to the Moon, the races are also an occasion for the Broome community to come together (figure 2.6). A further event that grew out of this community and represents its diversity to this day is *Shinju Matsuri*, the festival of the pearl. In the course of our stay, all of these events became significant places for my family and me to engage with community members and also become members of the community ourselves.

We had moved into Dora St in March 2012. Everything we owned fitted into the Holden Commodore Station wagon that Stephen and I, together with my friend Lars who was visiting from Germany, had driven over from Sydney. My partner Lina and our daughter Adele flew. Their arrival in the Kimberley was delayed for two days because of a cyclone that hit the Dampier coast near Karratha and saw flights from Perth to Broome abandoned. The kitchen was equipped. There was a washing machine, a double bed and sufficient cupboards, a dinner table with chairs, a sofa, and air-conditioning. We had brought the baby cot from Sydney, together with a ridiculous number of diapers, which not only turned out to be useful but also unnecessary because Broome is home to the two aforementioned shopping centers, one of which advertises itself with the slogan "Shopping in Paradise." We could have bought the very same baby cot at the local Target as we had brought from Sydney. But who knew?

As stated here, from our unit on Dora St we only had to cross the street to the old tamarind tree to greet Theresa Roe, one of Paddy Roe's daughters and the mother of Mr Roe (figure 2.7). At the time, she lived in the same rooms of the Old Hostel in which she had given birth to her eight children. Shortly after we moved in, my family and I visited Theresa, who sat on a white plastic chair, one of those you find all over the world. Stephen and I had seen her before, in October. "Stephen, yes," she exclaimed, "Good to see you . . . . Yeah, we're still fighting, like the Old Man [Paddy Roe, her father]. Keeping country strong."

She had repeated these sentences over and over in the past years. Many journalists, politicians, lawyers, campaigners, tourists and backpackers had passed by, interested in what the *Goolarabooloo* were doing—fighting the largest independent Australian oil and gas company and the Western Australian government. "Still fighting, like the Old Man. Keeping country strong," is not what I expected to hear. How was she fighting, sitting in this plastic chair near the tamarind tree?

Theresa, the matriarch of the Roe family, was cared for by her children. It was her choice to live in the Old Hostel for as long as she could. It was her way of saying, "I am still here to remind you what my father has done for this community." The surroundings of her block now offer social housing for many Indigenous families. The block accommodates *Mamabulanjin*, Broome Regional Aboriginal Medical Services (BRAMS) and a childcare facility inaugurated by Paddy Roe in 1987: *Yalagarr*, the first childcare center our daughter attended—and enjoyed, especially if her beloved carer Mio, who had migrated to Broome from Japan a few years before, was there.

Many of our other neighbors on Dora St were migrants from the Philippines, who worked in local tourism and other service industries as housekeepers or cooks. Obviously, the cultural diversity of Broome continued to grow. It was enriched by these new arrivals, who were eager to integrate into the community. This is reflected in the aforementioned *Shinju Matsuri* and its float parade, one of Broome's main transcultural tourist attractions. It was initiated by Japanese pearl divers to commemorate their dead, and has Sammy, a Chinese dragon, as its emblematic mascot, and the Filipino community—our neighbors—won the 2013 competition for best float design (figure 2.8).

The first time I sensed that we would live in one of these apartments on Dora St was when I visited Broome with Stephen in October 2011. The route we took to Notre Dame University was south, along Harmersley St, one of the main roads that passes the old Telegraph Building and Courthouse Markets, the police station, the prison, Our Lady Queen of Peace Cathedral and Matso's Restaurant. It then turns west and becomes Guy St. The third roundabout is where it crosses Dora St.

Thinking of roads, I remember driving out to Cable Beach for the first time and how I asked, "Why is this road not straight but has this long bend in it?" And Stephen replied, "Probably because old Paddy told them to build it this way. There might be some site in the dunes that needed to be avoided." Form Dora St into Broome and beyond, the Old Man had a significant impact on the infrastructure of the town—on how it is laid

Figure 2.7. Theresa and Phil Roe at Walmadany/James Price Point, 2013. Photo: Carsten Wergin.

out, on places that were to be preserved, and on places that were to be constructed. Living in the knot which is Dora St taught us this heritage and the legacy of the Roe family's entanglement with the town that Mr Roe was to preserve against the industrialization of Walmadany/James Price Point.

2.5 FAMILY AND FRIENDS

Amenities such as air-conditioned comfort in an extensive shopping center helped us to realize that this outback town was not so "outback" after all. However, Broome is isolated geographically—Derby, the next larger town, which has about 3,400 inhabitants—is 220 kilometers away. Otherwise, there is only the Indian Ocean and the country that surrounds it, which gave us good reason to feel claustrophobic at times.

As one cure for this, we learnt to make use of communal places to meet with friends and discuss the things that concerned the town on an informal basis, in particular of course the LNG proposal and the related conflict. The Courthouse Markets were held on every weekend and we often attended them on both Saturdays and Sundays (figure 2.9). We would meet Frans Hoogland, Jeanné Brown, or our dear friend the late Deb Vincent at their stalls to chat. Environs Kimberley had its own stall with information material. George H. would advertise his *Astro Tours*, and local musicians whom I knew from playing the saxophone with them would perform in front of the old Telegraph building, while our participation in the markets went to the point where my partner Lina sold some artisan products, which she had made herself.

Our integration into the community was in full swing. Many people had been generous without knowing us before, like those at *Mamabulanjin* and *Goolarri Media*. We did not want to disappoint them. To remain an impartial observer when you begin research through such ties is an impossible task. However, without them the research would never have been possible in the first place.

Again, this also had very practical reasons. As previously mentioned, housing prices in and around Broome had risen so significantly that we would not have found suitable accommodation without local support. In addition, the value of the Australian dollar was high because of the mining boom. This had kept the Australian economy strong throughout the GFC, but meant that the cost of one Australian dollar was between 75 and 80 Eurocents, which made it difficult for us to afford Broome.

My postdoc income was generous by European standards, but for a family of three with a monthly rent of about $2,000 AUS Dollar (around €1,500 EUR in 2013) for a two-bedroom unit, it quickly evaporated. What

Figure 2.8. Float of the Filipino Community during Shinju Matsuri 2013. Photo: Carsten Wergin.

we therefore needed in Broome, certainly more so than in other places, was to be part of its community, as its informal networks can compensate for economic constraints, while newly found friends can replace family and (childhood) friends who live thousands of kilometers away.

Behind the Aboriginal childcare and health facilities, Dora St ends and it meets Anne St, or as locals call it, the Broome Bronx. More members of the Roe family have houses in this area. They are respected and feared. Like the other branch of Paddy Roe's descendants, the Hunter family, they have a reputation for being proud and strong, and not backing away from a fight. Apart from physical violence, they suffer from alcohol and other addictions, heavy smoking, and bad diet. Members of the Roe family are thus easily classified as stereotypical Indigenous Australians: Their life expectancy lies somewhere in the mid-50s. Girls can become pregnant at a very young age.

Such stereotypes imply that Aboriginal people are dysfunctional within Settler-colonialist society. However, what I sought to emphasize was what they are capable of despite what they have to endure. How could a family that has been discredited as a bunch of dysfunctional, violent drunks stage a successful opposition movement against the interests of both the WA government and a multibillion dollar oil and gas company?

When we celebrated our farewell at the beach near Gantheaume Point, all of the Roe brothers were there: Phil, Ronnie and Mr Roe. "A rare moment," many of our friends attested. "You can always come back to us, you're family now, you always have a home in the Kimberley," is what Mr Roe told us that evening. This is the essence of the *Goolarabooloo* community: to provide for a place and people that welcome you, regardless of who you are.

It is not uncommon for friends to tell you at the end of a visit, "Come back soon, our door is always open for you." But would that mean the same thing? Or, likewise, would one welcome friends back a month, a week, a day after they had left? Would one let them stay unquestioned for a day, a week, a month longer? What Mr Roe said was, "You can come anytime, and you can stay forever."

Paddy Roe had taught his family to welcome others in this way, to offer a home for those who look for one. Parts of this welcoming attitude has found its way into the wider Broome community, whose informal networks made it possible for us to stay. Paddy Roe and his wife were in a similar situation when looking for a home and finding one with the Elders through whose country the *Lurujarri Heritage Trail* leads today. A detailed version of this narrative can be found in Paddy Roe and Stephen Muecke's publication *Reading the Country* (Benterrak, Mücke, and Roe 1984).

If you accept their hospitality, you have the chance (but not the obligation) to learn more from the *Goolarabooloo*, up to the point where people become part of the community themselves, like Frans Hoogland or Jeanné Brown. They begin to participate in daily activities, build their own homes on blocks of land at *Millibinyarri* or Sand St, and find ways to contribute to the community: they produce wood work, clapping sticks or boomerangs to be sold at the Courthouse Markets, or contribute music to the evening camp fire or scientific expertise to environmental assessment reports.

Frans Hoogland, the 70-something, always handsome Dutch man, lives at *Millibinyarri*/Coconut Wells in a hut he built himself out of sand and water. I think of how he told his arrival story: how he let go of everything, moved naked into the bush, under a tree, hardly eating or drinking, mostly meditating, until a guy got stuck with a Holden Ute. Hoogland helped him out, and in return got offered a beer. Afterward, he returned to town and back to civilization.

At the time, the bush that today is *Millibinyarri*/Coconut Wells block was rough, with no houses on it. Now there are roads that have been formed in the sand from the many car journeys in and through it. But bitumen you find only on the access track to the different private homes, mango farms and B&Bs that have opened left and right from the block, along the coastline, with beautiful views onto the Indian Ocean.

Figure 2.9. Courthouse Markets, Broome. Photo: Carsten Wergin.

By car, one has to drive out on Broome Rd toward Derby, which is the highway towards anywhere, really, as it is the only bitumen that leads into and out of town. To reach *Millibinyarri*/Coconut Wells, one has to take the famous turnoff at Cape Leveque Rd, which remains bituminized for another 20 minutes before turning into the dirt road that leads up the peninsula, which has also been sealed by now. This road had regularly been closed to traffic in the wet season because of flooding and will continue to be until the work is completed just past the Manari Rd turnoff. Fines for not obeying road closures are $1000 AUS Dollar per tyre (first rule: better not pack a spare).

Roughly five minutes past the Cape Leveque turnoff is the turnoff toward *Millibinyarri*/Coconut Wells, again a left. So first out of town, east, then north on Cape Leveque Rd, then west again, back toward the ocean. Along the coast runs the Song Cycle that stretches from One Arm Point at the northern tip of the Peninsula to *Bidjadanga*, some 40 kilometers south of Broome. The Trail covers parts of it, from *Minyirr*/Gantheaume Point in Broome to *Bindiangun*/Yellow River, some 15 kilometers north from Walmadany/James Price Point.

When driving out to visit Frans or others, I was often annoyed by the detour that the bitumen takes. However, as a Trail participant, I realized that if not for that detour, there would be no walk away from the lights and from *whitefella* business. The bitumen is built purposefully around the Song Cycle. *Goolarabooloo* know the short cuts through the bush—"proper" roads, not bitumen (Muecke 2007). Others will be obliged to take the detour.

No one knows this country like the *Goolarabooloo*, but the origins of the community are contested. Opponents argue that Paddy Roe invented a narrative about him being given custodianship from *Jabirr Jabirr* and *My-inirr/Djugun* Elders, even though there were enough youth around that such a transition would not have been necessary. Meanwhile, prior to the LNG proposal and subsequent conflict, *Jabirr Jabirr* and *Goolarabooloo* filed a joined claim over the area. The group split over the question of whether or not to industrialize the land and sea. As a result, in 2017, the *Bindunbur* Native Title determination did not recognize the *Goolarabooloo* as Traditional Owners despite of their knowledge, their close entanglement with "living country," their heritage legacy, and their unique position within the wider Broome community (chapter 6).

Another reason to find this determination wrong was given to me one afternoon when I went out to *Millibinyarri*/Coconut Wells to see Frans. My plan was to attend a meeting of *Goolarabooloo* families, in preparation for the three Trails that would happen in 2013. But the meeting was canceled. The three brothers had gotten into a fight the night before. Phil was still intoxicated and lay on the upper veranda of Mr Roe's house, near his mother Theresa and her daughter. Frans was with him. As I approached he quickly made his way down the stairs. He briefly explained that there was a problem with alcohol (again).

Instead of the meeting, I went with Frans to his place and sat on his veranda looking down toward the ocean. I was not prepared for another interview with him, so we spoke about general things, about how the Trail might develop, about potential outstations for tourists, about a couple who had parked their car at *Millibinyarri*/Coconut Wells and were currently walking the Trail by themselves. A man had been inspired during the 2012 Trail and wanted to take his partner. They had a tremendous amount of stuff—tent, cooking equipment and food. Frans was concerned about their whereabouts, since they did not have the option, like Trail participants, of a boat transfer at Willie Creek, but had to walk around it in order to stay on the Song Cycle.

The creek had a particular meaning for Frans. He once took me there in his little Suzuki 4WD. The tide was high. We saw a dolphin in the creek—a rather unusual site. I remembered the same experience from when I reached the end of my second walk with Phil for the report I prepared for the *Goolarabooloo*. I kept wondering how much the dolphin meant for me, as a companion and a guide. I googled, and came across an Aboriginal story of the *Burramadagal* clan of the *Dharrug* tribe called "The Dolphin Children."[9] It tells about three children who on an extremely hot day, despite the warning of their Elders, walked off, grew exhausted by the heat and eventually jumped from a cliff into what they thought was a billabong to cool down. Once they dived in they realized their mistake and found themselves being dragged out to sea. *Boomali*, a sea spirit, came to their rescue by turning them into dolphins. For the rest of their lives they could now play in the sea, but were never allowed to see their family again.

Both times I saw dolphins was on extremely hot days, and both times we were foolishly without enough water. So the dolphins seemed to indicate to us that it was time to turn around. However, that day Frans continued to the other side of the creek. He pointed to the Lore ground where he had experienced his first public part of a *Corroboree* at around Christmas time in the 1970s.

I had walked close to there. I had crossed this plain, which is covered with white salty sand from the floods during high tide. Mangroves sting your feet while you walk. There are hoofprints from the brumbies (wild horses) that I saw running in the distance when we descended from Willie Creek Pearl Farm back onto the Trail, onto the Song Cycle. I remember the goanna that had been run over and was tucked away safely in the back of a four-wheel driv. I never saw another piece of it. Richard Hunter, the storyteller, must have eaten it in a quiet moment. The Hunter.

"Now you have heard quite enough, no?" I remember him asking after an interview I conducted with him. I think of Marshall Sahlins (1972) and how he cherished the freedom of those hunters, those supposed primitives who are not caught in the economy of apparent imminent lack. Richard wanted to be out on country with nothing but his fishing gear. Repeatedly, people who accompanied him exclaimed in astonishment at how for hours he would sit out near the ocean trying to catch a fish, a crab, a turtle—for the camera, for another media campaign, to create awareness of the region and what was threatening it. The thought of him generating awareness for the local environment, and not shying away from media-sensitive "staged authenticities" to do so, brought me back to reality.

There is endless commitment reflected in these actions of Frans, and Richard, and Mr Roe. Who dares not to recognize such "traditional ownership?" For reasons that will become apparent in the following chapters, this question brings me to the last knot I wish to mention here: the NGO Environs Kimberley.

2.6 ENVIRONS KIMBERLEY: AN INSIDER'S VIEW ON MULTISITED PROTEST

Environs Kimberley (EK) was initially set up in 1996 to support Traditional Owners (TOs) in protecting the Kimberley from the damming of the Fitzroy River and large-scale land clearing to grow cotton. Their scientists and other staff work with communities, landholders, and land managers, and especially TOs and their ranger groups, on issues such as wildfire, weeds, and feral animals to protect local plants, animals, and their habitats.[10]

Martin Pritchard, the director of EK, took up his position just before the Walmadany/James Price Point conflict took off in full force. I remember the conversation I had with him about the collaborative powers of a protest movement with no defined hierarchical structures or other top-down thinking involved. As he explained, EK had a decisive role in them:

> *Martin Pritchard:* I saw the job advertised, and while I was writing my application, there was the [ABC] *Four Corners* [TV] program on the gas [plant], which made me redouble my efforts (. . .) I started on December 5th 2008. That was two weeks before Colin Barnett announced James Price Point as the site.
>
> *Carsten Wergin:* What motivated you when you saw that *Four Corners* program?
>
> *MP:* I'd always had an interest in the Kimberley. Didn't know much about it other than that it was this incredible natural and cultural landscape. That it was big, huge, and sounded exciting. When I saw the gas issue I thought I'd really like to be involved in the campaigning side of it, because I could see that (. . .) it had the capacity to become something that was national, to a lesser extent, an international issue (. . .)
>
> *Carsten Wergin:* What did your work look like when you arrived?
>
> *MP:* I was taking over from Maria Mann who virtually, with the help of others, set up the organization. It was pretty daunting starting the work, given that she'd been in here for twelve years and had built up an incredible knowledge of a whole variety of different things. Then, a two-week handover, and at the end of the hand-over, Colin Barnett was saying James Price Point (. . .) I tried to go through all the work that was on the hard drive, talking to the board and as many different people in the community as I could about what the issues were and what people's sentiments were . . . . Connecting with people like the KLC, other individual people like [Mr Roe]. People like Anne Poelina, Mitch Torres (. . .) Two weeks after I started, Barnett said James Price Point. He said something like, "We're really willing to talk to people in Broome about this." So I immediately rang up the ABC when he was talking and challenged him to a meeting. Within about a month of starting, I'd done quite a few media entries about the issue and had a meeting with the Premier in my first month. (Transcript 130513_1_MP)

Martin Pritchard was thrown into the conflict and had to take charge. He entered from the outside and immediately invested in getting to know the community and its main concerns about the industrialization proposal. Significant new collaborations began, and EK was their catalyst:

Figure 2.10. Richard Hunter Telling about Country. Photo: Lina Pranaitytė-Wergin.

Martin Pritchard: Because I was paid to work we were pushing (. . .) the story from Broome in those times (. . .) We had an [EK] No Gas campaign strategy that was done with someone who was experienced in international campaigning. We had a really clear way forward that we were going. We had five main parts to our strategy, and one of them was to support the Broome community (. . .) We provided Non-Violent Direct Action [NVDA] training (. . .) I had visits from the Office of State Security, which is basically law enforcement, but they are like a special squad that deal with VIPs . . . . They'll be with the Premier, travelling around . . .

Carsten Wergin: They came here?

MP: Yes, they came to see me [laughs] and said, "Where do you think this is going to go?" I said, "Well, do you remember the Franklin [dam protest]?" They went, "Ah, yes." "Well, I expect lots and lots of people to be coming here to protest, and yes, it's going to be pretty tough." (. . .) We could see a need to provide workshops on [NVDA] so that people have an avenue that was non-violent (. . .) There's lots of websites from Australia, particularly from the US, about the principles of [NVDA]. Some of it comes from the principles of Mahatma Gandhi and other people who've practiced non-violence. It ranges from writing a letter, to locking onto a vehicle. There's a list of over a hundred things that you can do that's [NVDA]. It's not just about civil disobedience. It's about direct action. Out from that, there was further community discussion, and then we got to a point where part of the assessment had a social side to it, social impacts (. . .) People saw that we'd been ramping up the environmental concerns, but there wasn't really a body that could talk about social impacts. [EK] could, but that's not where their expertise lies.

[So] people wouldn't necessarily recognize us as legitimate commentators. (Transcript 130513_1_MP)

Media attention grew considerably with the public announcement of Walmadany/James Price Point as the site on which the LNG precinct was to be constructed. Gradually, the protest expanded from focusing on environmental issues to addressing social concerns, too. NVDA workshops generated additional momentum as they stimulated creative engagement by the community. People were getting more involved, and they brought more expertise, but also made EK recognize its limits as a predominantly environmental NGO. As a result, protest actions became truly multisited in an ethnographic sense: not only because they happened in different places such as Broome, "on country," in courtrooms, or at music events in Byron Bay or Melbourne but also because protests were fought from different angles, incorporating environmental, social, economic, and Indigenous concerns.

Martin Pritchard: As people began to realize that social impacts were a problem (. . .) we decided: Let's have a bit of a forum where we can talk about and broadcast those concerns about social impacts. That's where the Broome Community No Gas Campaign came from (. . .)

Carsten Wergin: That began in 2009?

MP: 2009, 2010, yes. It was quite amazing. I don't know if it was the first or second workshop where someone brought up the idea of lock-ons. I think it was Gerard and a few other people . . . *Goolarabooloo* immediately objected, "No way. You crazy? That's mad." Shaun brought us the first lock-on (. . .) That was the first lock-on in the Kimberley! Next minute there was people making their own lock-ons in their back sheds and taking them up to different parts of the community. Just dropping them off going, "Here, I've got this design off the net." (Transcript 130513_1_MP)

The creativity that the community involvement generated helped the protest to develop its full transecological force, as social and environmental concerns began to link up with "cultural" concerns brought forth by the *Goolarabooloo*:

Martin Pritchard: From a relatively small base of concerned people, it built up to photographers, filmmakers, community workers, lawyers, doctors, nurses, teachers, traders. . . . Just a big build-up of people's understanding and knowledge of campaigning and people taking their own initiatives to do things (. . .) We got to a point where we had managed to get the most support that we possibly could in Broome from people who would be concerned about the environmental impacts (. . .) [With the social impacts we] got those other people who were possibly sitting on the fence, or not supporting the gas but not particularly motivated to take action, to start becoming more activated and engaged (. . .) In terms of *Goolarabooloo* interests, [Mr Roe] went to court on a couple of occasions. It seemed like they had a very targeted strategy with their court cases. It wasn't necessarily about media and public knowledge around it.

I guess what started to build up was things like [Mr Roe]'s profile and Albert Wiggan when they did a tour with John Butler around the country. I mean concerts. It was maybe around 20,000 people in Melbourne (. . .) John gets on stage and he'd introduce [Mr Roe] and talk about what was being proposed here. [Mr Roe] would talk about his country and the support he needed (. . .) They might have done something at the Byron Bay Blues Festival and they might have done something on John's tour. That raised the profile to a quite prominent degree. Then in Broome you had the *Walmadany* concert on Male Oval that the Pigrams organized. It was called the Heritage Concert. There was about 5,000 people that went on Male Oval to that concert. Even though it wasn't billed as anti-gas it was quite well understood what it was about: protection of *Walmadany* and heritage. That's another thing that raised the profile significantly. Then people do stuff about the Trail. People who are connected to *Goolarabooloo*. Some people come back, people in the community helping them put together a . . . submission under a national heritages scheme . . . meeting with ministers... that kind of stuff. I guess it was a bit of a mixture of ways that the cultural interest got boosted up there alongside the environmental interests and the social interests. (Transcript 130513_1_MP)

Martin Pritchard's personal involvement and commitment led him and EK from the recognition of environmental concerns, via the social, to the cultural. This was inevitable when working with the *Goolarabooloo*, since the three dimensions are closely entangled in "living country" and living heritage. The protest brought the values of "living country" to national attention and gave voice to its Australian Dreaming through conventional media and cross-country networks of people who are concerned about the environment and willing to organize in its defence. This brought the message about the conflict over Walmadany/James Price Point to Perth, and via Melbourne to Tasmania, which has made its own significant contribution to the history of environmental movements with the aforementioned successful protest against the construction of the Franklin dam in the 1980s.

All the collaborations that Martin Pritchard touches upon in the above interview excerpts are fleshed out in more detail in the following chapters. At this point, his account is useful as an introduction to their entanglements and also as a reminder of what I have argued thus far: that this ethnography is about resource assemblages and the ruptures they generate, about events that are "toxic" for late liberalism but that allow for modest biocultural hope brought about in and through lived experience. This allows this inquiry to present glimpses of an Australian Dreaming, of a world in collaboration and a viable alternative to the late-liberal mainstream.

NOTES

1. https://www.yawuru.com/our-organisation/nyamba-buru-yawuru-about/ (Accessed November 9, 2021).

2. This experience also sensitized me to the general mistrust of anthropologists that is held by Indigenous people in Australia, which is due to the very history of the discipline. Among others, anthropology has been a significant contributor of allegedly scientific evidence for the classification of Indigenous people as "primitive" and inferior to Settler-colonialists (Wergin and Erckenbrecht 2018).

3. To be fair, the occasional snake or scorpion popped up in the vicinity of the camp. But the animals usually see people before people see them, and avoid crossing their paths.

4. Personal communication with the associate.

5. Further information is accessible via the URL: http://museum.wa.gov.au/newmuseum (Accessed November 21, 2021).

6. Personal email exchanges with members of the group.

7. Contributions are accessible via the URL: http://www.kimberleyeconomicforum.com.au/presentations.html (Accessed November 12, 2021).

8. *Bunuba* Elder Jimmy "Dillon" Andrews is a custodian of this legendary tale and shares it with tourists in a specially designed tour. The tour is also used to train and employ young *Bunuba* people in the tourism business.

9. Source: http://www.artistwd.com/joyzine/australia/dreaming/dolphin_children.php (Date retrieved December 1, 2021).

10. Further information is accessible via the URL: https://www.environskimberley.org.au (Accessed November 9, 2021).

Chapter 3

From Transculturality to Transecology

The Environmental Humanities offers a new means of engaging with Australia's transcultural past. I use the term "transcultural" here in its most general sense, to emphasize processuality as the very foundation of cultural expression (Burke 2009; Coronil 1997; Ortiz 1995; Welsch 1999). Multicultural and intercultural exchanges falsely assume that there are cultural entities to begin with. In fact, the opposite is the case: cultural entities are not the origin of cultural exchange processes, but the imaginary results of them. More often than not, they are constructed to exercise power over other contributors to these processes, be they human or more-than-human.

With regard to such exercises of power in and through discourse, the Kimberley region and its tourism industry present a particular case in point. For all its strong Aboriginal links, the visible presence of Indigenous views in society, culture, and planning is only gradually evolving. Meanwhile, the "paradox of visiting Aboriginal Australia" (Muecke 1999, 54) still sees many popular images of the Kimberley and other remote regions, such as that of a pristine landscape, wilderness or "last frontier," ignore the original and continued presence of Aboriginal people.

In other instances, Indigenous Australians are confined to the conventional realms of exotic attractions and producers of artifacts. These tourism imaginaries transport the common ideal of a multicultural Australia as described above. In the "Multicultural Affairs Policy" of 2017, the then Prime Minister Malcolm Turnbull emphasized, "The glue that holds us together is mutual respect—a deep recognition that each of us is entitled to the same respect, the same dignity, the same opportunities."[1] Turnbull adopts the conventional understanding of multiculturalism and related "same but different" politics in his government policy. Australians share the same rights and obligations as citizens but different cultural roots. This approach, however, is far from the reality in the Kimberley, where not only have Indigenous and non-Indigenous people traditionally been mixed, but diverse Indigenous groups, despite their different languages, clan, and skin groups, have also shared their culture and traditions in ceremony and custom "since day one."[2]

The Kimberley is therefore emblematic of what I term a "Transcultural Australia." Its foundations are well beyond the common agreement on mutual respect advocated through government policy, because exchange processes are at the very roots of its cultural expression. The theoretical landscape of this book takes this processuality as its starting point, while I contend that the values of Transcultural Australia will be accounted for in many more ethnographic studies to come.

I draw on theory that links up with the notions of a "living country" and living heritage outlined above to examine the conflict over Walmadany/James Price Point as a means by which the ecological and transcultural values of the Kimberley can be presented differently: as biocultural entanglements and rich sources for "multispecies storytelling" (van Dooren 2014). To take this diversity of world(view)s, multiple species, and their onto-epistemic partnerships seriously, the discussion needs to begin with the assessment of the values attributed to the land. In that they advocate for the extraction of resources without prior attention given to the values already in place, industrialization proposals run against the premise that landscapes have intrinsic value.

Such large-scale exploration of natural resources has had devastating effects in all parts of the world as it continues to severely undermine local traditions and customs, culture and heritage (Swyngedouw 2010). The reason for this is that industrialization projects downplay or ignore the intrinsic values of a place that are linked to ethical concerns and preserved in localized tradition and custom. Meanwhile, the age of the Anthropocene is also the age of a growing awareness of related socio-political and economic inequalities that are the result of human-made environmental degradation (Blühdorn and Welsh 2008).

The conflict over Walmadany/James Price Point is another expression of this growing awareness, and its collaborative work of Indigenous and non-Indigenous groups can be a source of policies that create alternative pathways toward a future beyond the resources boom. In this sense, my ethnographic analysis ties in with a wider interest from scholars who seek to write anthropologies of the "otherwise" (Povinelli 2002; Holbraad, Pedersen, and Viveiros de Castro 2014; Holbraad and Pedersen 2017). I have adopted this perspective of the potential "otherwise" because it offers new means by which to acknowledge multiple world(view)s and their onto-epistemic partnerships—or, to formulate my motivation for this project in a poignant question: If the Anthropocene teaches us "how to die" (Scranton 2013), what can the transcultural heritage and ecology of the Kimberley teach us about how it might be otherwise?

My ethnography demonstrates how the diverse regimes of value attributed to a particular area link up with the wider Dampier Peninsula, for example in successful Indigenous tourism experiences that offer further means by which to recognize and protect the "vital materialism" of Australia's transcultural past and present (Bennett 2010). I term these values equally vital because they are preserved in Indigenous Lore and custom and transmitted through cultural experiences such as the *Lurujarri Heritage Trail*. These values offer participants a concrete means by which to do and think differently about Australia as a whole, not as a multicultural cacophony under the same law, but as a transcultural lifeworld that recognizes both Western law and Aboriginal Lore.

On the following pages, such an onto-epistemic partnership between law and Lore becomes visible in ethnographic examples of emblematic heritage products of the Kimberley that were put to use in order to counter the industrialization proposal for Walmadany/James Price Point. As a result, it becomes possible and necessary to reassess transculturality in ecological terms, or to coin a term, as "transecology." By introducing this theoretical notion of "transecology," I seek to make an ethnographically informed contribution to the larger, future-oriented, interdisciplinary program that is the Environmental Humanities and that produces "thick description" beyond the mere "interpretation of cultures" (Geertz 1973). To do so, I draw on the Environmental Humanities and other social science research, as I draw on Indigenous knowledge and practice.

As the case studies below demonstrate, both are equally important sources for research and ethical standards that depart from the Modernist project (Hulme 2011). Even more so are the collaborations that emerge from participation in the *Lurujarri Heritage Trail*, which are exemplary of an amalgamation that makes for what I term "hidden quality of Indigenous tourism experiences."

Central to my discussion is how these qualities help to redefine human/non-human relations not as separate but entangled, and therefore in radically old/new ways. They are new ways for two reasons: first, they lead beyond the search for difference toward collaboration and commitment, and second, they lead beyond the teleological-minded futurism of modernity toward the acknowledgment of what is in place. Related experiences are thus equally old and new, because they reintroduce Aboriginal Lore and culture as means of engaging with the land. They draw on how it has been "since day one," not to criticize what has been lost in the "contact situation" but to draw on "living country" as "living heritage" to envision a common future for Australia—from contact to collaboration.

3.1 HERITAGE IN THE ANTHROPOCENE

In order to combat social exclusion, there remains a need to introduce the voices of people and places in so-called peripheral regions into debates on global transformations (Altman 2010; Chabay 2010; in't Veld 2010; Lea 2006; Taylor-Gooby 2013). These people and places provide for more than just a different representation of the world. They demand the acknowledgment of different realities or "worlds" in themselves as the precondition of decolonization.

My ethnography of the collaborations that emerged in order to preserve Walmadany/James Price Point from industrial development problematizes the difficulties that come with such acknowledgment. A significant part of those difficulties stem from failed attempts to translate between different world(view)s. Translations are necessary for diverse collaborations to work, but equally dangerous because of their tendency to override the world(view) translated with the one it is translated into. This can lead to the subjection, or even loss, of one worldview for the benefit of the other.

The notion of transculturality is helpful here because of its emphasis on processuality rather than stasis—on networks and assemblages, rather than entities and separations. However, with regard to the context in which this ethnography is set, I found that the concept of transculturality did not allow me to give the particularities of the local environment, its "living country" and living heritage, sufficient recognition. I therefore introduced the notion of "transecology" to better describe the diverse collaborations between human and more-than-human actors that I observed. In what follows, I describe how it came about, triggered by the first interview I conducted during fieldwork.

It was May 2012, and I sat with Elisabeth F. at her recently opened five-star hotel Pinctada Cable Beach Resort and Spa in Broome. We met a few days after she had contributed to the Kimberley Economic Forum held at Cable Beach Club under the title "Owning the Future." I was impressed by her talk, which for me reached further than the late-liberal analysis that governed most other contributions. The latter concentrated on suggestions for how to best seek economic opportunities, which mainly concerned imaginaries of an expanding resource industry in the Kimberley region.

Elisabeth F. spoke about values located in the place, which she elaborates on in the excerpt below. The account presented here illustrates why a move from a transcultural to a transecological perspective becomes necessary in order to illustrate the values at play in this particular region. In it, she sets out from her views on the differences between Indigenous and non-Indigenous culture, but ends on how important it has been for her to incorporate environmental differences, as well:

> The Indigenous culture is a completely different one. It hasn't been . . . what we call "industrialized" in any way. So, you can't assume that coming from what we call a "primitive" life[style]—what we *call* primitive—a non-industrial lifestyle. . . . You cannot suddenly transform into a member of the community who does everything that everybody else does: goes to work [at] nine o'clock, goes home at five, shops in the supermarket, cooks dinner and pays a mortgage. You're talking about completely different . . . (PAUSE) . . . cultures. . . . So separate that the bridge has to be a very long bridge to get from one to the other. And that's what's happening now. There's that very long bridge that people are now walking on; that will go from (. . .) this very isolated, very strong [culture], cultural[ly] and Lore-wise. . . . Which was completely and absolutely misunderstood (. . .) and totally negated by the white man (. . .) This long, long journey from something that actually did work and was highly, highly sophisticated, and very strong culturally, to what we know . . . which is really very simple by comparison. That's a massive journey. And originally, there was a journey being demanded [by Settler-colonialists] that none of that had a value (. . .) But not only that: people will be displaced unless they bring their culture forward with them. And this is the critical thing. . . . It's once again those values: What's important to you? What do you need to take forward with you in life that makes you feel that you have a… (PAUSE)… That you have a backbone? That you have a substance that you can rely on? A core, if you like.
>
> So we're nowhere near this path over here (. . .) The original dream (*sic*) was, "Forget all that. Just come here. You're very primitive. You won't be able to do anything anyway, so we'll do this. You'll always be basic," which was totally inappropriate (. . .) Up in the Northern Territory, for example, in Arnhem Land, they [Aboriginal people] *traded* (. . .) with the Macassans and everybody for generations (. . .) They're very independent. They were healthy. Beautiful, rich land (. . .) They had cattle. . . . In Kakadu, even though it's a national park, they still retain cattle there. That's for use purely by the Traditional Owners (. . .) So they've been mixing with Asia for a long time. Their mixing with white man was a new thing. You know, the path from being the nomadic, tribal person, tribal culture that had no possessions . . . to our lifestyle . . . (PAUSE) . . . is a big step (. . .) People are still making that journey. And it became a very political thing when land rights came in: The much greater understanding, much insistence on that. The *Yawuru* people only recently, last year, May the 14th—were actually given title to their land. May the 14th, 2011. (Transcript 120504_MaPa)

Up to this point, Elisabeth F.'s account makes itself readily available for transcultural analysis. Based on her long-term experiences in the Kimberley, she focuses on the journey that the different groups that met in the colonial situation, the Indigenous and the non-Indigenous, still have in front of them. In her view, their relationship was falsely described in hierarchical terms that expected the alleged "primitive" to catch up with Western civilization. Instead, she argues, this history needs to be rewritten in order to acknowledge the values of an Aboriginal culture that might be more sophisticated, from a contemporary point of view, than the Modernist lifestyle most people nowadays adhere to.

In the interview, she then gradually moved from this transcultural view towards what I conceive to be a transecological one:

> *Elisabeth F.:* Maybe [the hippies] were trying to go the other way. They were looking for whatever it was that the Aboriginal community might offer them. Just the other way around—without possessions and, rather, this connection to country (. . .) And that is actually what a lot of people here do aspire to. And I think that's where a lot of the objection for the gas [plant] comes in. I think there are different levels happening. Some are professional protesters who are brought in from interstate to be very, very loud (. . .) And then there's the local community, some of whom are truly passionate about this place and about the coastline, and whose main motivation in disagreement is a genuine fear that things will be destroyed. Not so much that they will change, but that they will actually be destroyed (. . .)
>
> *Carsten Wergin:* I spoke with Frans Hoogland, who has lived with the *Goolarabooloo* since the 70s. [He is] from Holland, originally. He doesn't talk like somebody who has any monetary interest in fighting anything. But he has a genuine perception that this needs to be protected.
>
> *EF:* Yes. And I believe that is a valid position. It shouldn't be shouted down, shouldn't be seen as being terrible. It should be recognized, as is every opinion, because there is value in it. Progress today really should be done to such a standard that it recognizes how precious this area is. And simple is not good enough. What we already know is not good [enough] (. . .) I use this hotel as an example. When I built this, it was very important to me that I built something that reflected Broome. I didn't want to be Bali. I didn't want to be Melbourne (. . .) I wanted this to reflect Broome, and so I built it to actually *recognize* the climate. So, from now on [early May], you don't need air conditioners in this property, really. All the doors are open in the restaurant even right through November, December. The only time we close them is, in fact, because of the rain. So, everything is built so that you get this amazing breeze. All the rooms are the same. You open your doors. You've got your fly screens. You open your bathroom louvers, fabulous breeze (. . .) All over, it's beautiful. The spa is not little, closed-in rooms that have no light. They all have louvers. So, this time of the year, you lie there, and you can hear the birds, and you can hear the rustling of the wind and

> the breeze in the leaves. And you feel the breeze across your body, and it's beautiful. I mean it's beautiful! To me, that's celebrating where we are (. . .) And then also recognizing that this is not a tropical climate. This is actually a semiarid climate. We have a very shallow water table that turns to salinity very quickly. So, our landscaping was very much based on understanding that and using, incorporating plants that had low water usage, but also that were a reflection of this area (. . .)
>
> Now, a lot of these things cost an enormous amount of money. And one would say, "Why the hell did you do it?" It's a very good question because it's very hard to recover in hospitality. But, to me, it was a sense of responsibility to a community and to a place. Now, if I can have that with my short pockets, these mining towns, mining companies could certainly have it with their deep pockets. I mean they have a social, moral responsibility to this community and to this country (. . .) And that doesn't say that they can't do what they're going to do. That doesn't say that they can't harvest their minerals. It doesn't say that at all. I'm a pro-development person. What it does say is, "Have I heard the very best answer yet from you? Is really that the very, very, very best solution that can be found for this?" (Transcript 120504_MaPa)

The argument made by Elisabeth F. in the above interview excerpt slowly moves from a transcultural world(view) to a transecological one. She moves from the differences between Indigenous and non-Indigenous culture, the violence of Settler-colonialism, the trade-routes between Australia and Asia, the cattle herding of Indigenous people in the Northern Territory, to the particularities of the local environment and the need to recognize them, to develop with rather than despise them. She describes why these particularities are valuable in their own right, how they add value to her hotel, and how acknowledging them is a responsibility for everyone, but especially for those who have the means to transform a place—entrepreneurs who build hotels and, last but not least, mining companies.

This argument put forward by Elisabeth F. is what motivated me to think about my fieldsite in transecological rather than transcultural terms. Though she does not mention "living country," let alone *Bugarrigarra*, she emphasizes a responsibility for the land, for values that are in place, which is close to what Aboriginal Lore and culture proffer: the interrelationship of people and place, a reciprocal, *care-full* acknowledgment of interdependent responsibilities (Wergin 2021). To put it differently, the notion of transculturality opens up a theoretical toolkit with which to acknowledge processuality as that which precedes cultural entities. The term transecology adds the ecological dimension to this acknowledgment of processuality.

Transecology recognizes that "living country" demands for experience as the only means through which objectification of its living heritage is possible, if such objectification is not to override the fragile lifeworld that unfolds through engagement with it but to preserve it. Our actions have consequences not only for us but also for a multitude of other species, people, and place, human and more-than-human. While this is a commonplace, the resources industry is a prime example of how Modernist ideologies of progress and development continuously lead people to forget this transecological entanglement.

The Australian Dreaming accommodates those who warn of the predicament of the Anthropocene, and who are motivated by the legitimate urge to create awareness of the consequences of human interventionism beyond humanity, no matter whether the impact is immediately discernible or not. Related scholarship, notably that of eco-feminists, fosters the search for more "mutualistic forms of rationality" (Plumwood 2009, 116). From this perspective, the Anthropocene is a self-fulfilling prophecy, as environmental philosopher Val Plumwood explains, "When we hyperseparate ourselves from nature and reduce it conceptually, we not only lose the ability to empathize and to see the non-human sphere in ethical terms but also get a false sense of our own character and location that includes an illusory sense of agency and autonomy" (ibid., 117). In line with Plumwood's critique of the "hyperseparation" of humans and nature, the transecological approach shifts the conventional focus on the consequences of human interventionism toward ecological factors that are specific to a place. Their acknowledgment is necessary in order to provide "the best possible answer," as Elisabeth F. has attempted in the design of her hotel. The Australian Dreaming thus leaves the realm of the transcultural. As a transecological project it does away with hyperseparations on the level of alleged cultural and ecological difference.

Transecology adds to the transcultural project those spheres in which the term "culture" departs from human-centeredness. It draws on contributions from the above-mentioned eco-feminists, experimental philosophers and other scholars, notably those in the Environmental Humanities, all of whom share an understanding of people and culture(s) as intrinsic to nature. Viewed through this transecological lens, the Australian Dreaming leaves the dualism of "culture (human) versus nature (non-human)" behind as only another reductionist concept that adds to hyperseparations already in place.[3]

To be clear, my criticism is that the differentiations that binarisms generate are much too often elevated to, or falsely represented as, dualisms in competition with each other. The opposition movement to the industrialization proposal for Walmadany/James Price Point is a case in point. It lends itself to an interpretation of "us versus them." However, the transecological approach will demonstrate that it did not prevail against government and industry because it was "better than them." Instead, it prevailed because it generated multifaceted collaboration grounded in "living country."

Modernist approaches offer no intellectual means to adequately analyze the emergence of these collaborations—or, as Plumwood writes, "modernity's philosophical contribution (. . .) kills off the driver without questioning the reduced concept of materiality that was its other side" (Plumwood 2009, 119). Here, the concept of transecology opens up a theoretical space in which to acknowledge collaborations as the sources of what are commonly perceived to be separate world(view)s. This is why it is not separate world(view)s played out against each other, but the world-making activities of diverse collaborators that are at the heart of my ethnographic engagement.

Rather than focusing on the representations, translations, and other anthropomorphisms inflicted on people and place, their heritage and culture, the transecological approach seeks to enable the latter to speak for themselves through the (ethnographic) description of their co-becoming. This is not to say that mainstream economics, government institutions, courts of law, or supranational institutions such as UNESCO would not have a significant impact on such processes of co-becoming, despite the fact that their decision-making cannot do without representationism, as they need to approach what is at issue in quantifiable terms. Meanwhile, the transecological approach reminds us to shift the (ethnographic) focus away from related representationisms. This is because, with transecology in mind, agential realism like the successful staging of the opposition movement against the industrialization of Walmadany/James Price Point are of more fundamental relevance. They produce imaginaries of how it might be "otherwise"—the coming-into-being of an Australian Dreaming after the resources boom.

So far, I have proposed transecology as a theoretical means to recognize:

(a) processuality as the very foundation of world-making activities (much like the notion of transculturality), and
(b) human and more-than-human collaborators as onto-epistemic partners in these activities (extending the notion of transculturality into the realm of the more-than-human).

The notion of transecology advances the transcultural project beyond its doublebind, which is that the "nature" of transculturality remains unrepresentable. Transculturality runs a risk of promoting the very dualism that transecology seeks to overcome—that of humans speaking for nature. Meanwhile, transecology focuses attention on the agential realism that bring world(view)s into being. In this sense, "nature" is woven into the notion of transecology not only as that which can be discovered but also more importantly, as that which makes the discovery possible in the first place (cf. Neimanis 2015). As such, a critical engagement with more-than-human actors that anthropologist Donna Haraway sums up in the formulation to "make kin" becomes possible (Haraway 2015, 161).

The question that needs to be addressed then is an ethical one, because multispecies stories, and with them a theory of transecology (cannot but) advocate a relational ethical approach that "multiplies perspectives on what counts as 'the good'" (van Dooren, Kirksey, and Münster 2016, 15). The central concern in this approach is not the survival of humanity. It advocates instead environmental decision-making that is collective in the sense that it acknowledges the co-becoming of an eco-ethological multitude. Possible extinction (including human extinction) is not the exception but the norm. This poses a problem for the social sciences and humanities, and anthropology in particular, a discipline dedicated to all that is and makes us human.

While from a mainstream perspective the acceptance of possible human extinction might seem radical—even contradictory to the very purpose of scientific enquiry—Indigenous views of the world (and not only in Australia) have shared this acceptance "since day one." The resulting conflict points to the very foundations of Settler-colonialism: its four pillars of dualisms, linear time, ownership, and naming (chapter 4). These pillars, on which Settler-colonialism rests, are categorically opposed to a world perceived in intra-action. The reason for this is that the pillars generate a decision-making dependency on representations of reality, on values attributed to reality, on names given to people and places, on entitlements and on a processuality that is teleological and not immanent, not dependent on the "here and now," on lived experience.

Because of these dependencies on Modernist constraints, a constructive engagement with how the predicament of the Anthropocene is understood and approached today is tied to the wider project of decolonization. A transecological perspective makes this obvious, because it shows that a planetary future will neither be determined by humans alone nor within the realms of representationism. It needs the recognition of alternative world(view)s. Political philosopher Benjamin Kunkel (2017, 6) adds, "the question of modern humanity's past and future ecological trajectory can't be intelligently posed except as a question about capitalism."

Capitalism has been accepted as a main cause of the contemporary global ecological crisis. Kunkel therefore asks for a critical engagement with the disease. But how can there be a remedy found within it? To answer this question, the transecological approach proposes that research(ers) engage with what exists despite capitalist constraints and the impact of Settler-colonialism and neoliberalism on the fragile life-worlds of Indigenous people. In other words, it contends that a cure is to be found elsewhere: in stories of the "otherwise" that highlight alternative world(view)s of multispecies entanglement.

Fulfilling the demands of late liberalism and casting themselves as a "real native" might be appealing to some Indigenous Australians because of a false promise of equality and access to resources. This promise is again played out on the four pillars of Settler-colonialism. Many Indigenous groups continue to ascribe themselves to Settler-colonialist ideals because they see no alternative other than to accept its pillars and the means of representation that they ask for, such as that of the "real native" and other "self-exotisations" (Wergin 2010a, 2015). As Povinelli explains, "hegemonic domination [. . .] works primarily by inspiring in the indigenous [*sic*] subject a desire to identify with an impossible form of being" (Povinelli 1998, 27).

The wider concerns that researchers of Settler-colonialist Australia are raising are linked to the continued exercise of "biopolitics" over its first people (Foucault 1978). In *Geontologies: A requiem to late liberalism* (2016), Povinelli draws on ethnographic examples from her work with the *Karrabing Aboriginal Film Collective* to show how late-liberal Australia bases its powers on related claims to anticipate, protect, and enhance "the biological and psychological needs, wants, and desires of its citizens" (ibid., 3). Related to this is my attempt, using the transecological approach, to relativize Modernist constraints and regimes of value within a wider co-becoming. To put it differently, with the end of "Cheap Nature", the very sources of potential profit as a driving factor in Settler-colonialist Australia no longer exist. What will be viable alternatives? To answer this question, the transecological approach emphasizes "the world in which these enclosures are no longer, or have never been, relevant, sensible, or practical" (ibid., 16).

Viewed through the transecological lens, the *Lurujarri Heritage Trail* is a case in point, while the conflict over the proposed LNG facility at Walmadany/James Price Point can be termed yet another catalyst for the problems faced by late liberalism, next to what Povinelli terms the Desert, the Animist, and the Virus (ibid.). Envisioning participation in the *Lurujarri Heritage Trail* might not invoke the same sentiments as envisioning oneself faced with the Desert, the Animist, or the Virus. However, as argued above, its lived experience can be the cause of substantial "disturbances" for the late-liberal project, because the Trail raises or enhances doubts within its participants about late liberalism's ability to continue to anticipate, protect, and enhance the world.[4]

Meanwhile, Native Title has arguably failed to incorporate Indigenous Lore and culture into Settler-colonialist legislation. However, this institutional failure remains of crucial significance, as it marks another political and conceptual "disturbance." This time, the disturbance is caused because of the mere attempt to work through onto-epistemic differences, which is what Native Title legislation seeks to do, and the failures of which raise legitimate doubts over the ability of the Australian late-liberal mainstream to cater for all its citizens.

Transecological theory and practice equally draw on such "disturbances." They seek out "toxic" events, such as protest actions, and the uncertainties that stem from them, as glimpses of sovereignty and into the "otherwise" (Povinelli 2016, 27). Transecological studies embrace "toxic" events, such as the protests against the LNG facility at Walmadany/James Price Point, because what might be toxic for late liberalism might also carry biocultural hope for those who already are at their existential limits. This adds ethical quality to the transecological approach, despite the fact that it relativizes possible human extinction, as thus far:

> The barely perceptible but intense daily struggles of many people to remain in the realm of the extreme poor rather than slip into something worse (. . .) only lightly scratch the retina of dominant ethical and political discourse because the effort of endurance and its incredible creative energy appears as nothing, laziness, sloth, and the unchanging. (Povinelli 2016, 21–2)

The transecological approach highlights people, places and also Indigenous tourism experiences such as the *Lurujarri Heritage Trail* as sound evidence for such "efforts of endurance" in Settler-colonialist Australia. More often than not, they are overshadowed by the stereotypical appearance of "nothing, laziness, sloth, and the unchanging." The realization of a Swiss backpacker that "not all Aboriginal people are drunks" is a poignant case in point. Her direct encounter with Indigenous hosts who shared their knowledge about country might equally be described in Povinelli's terms as a source of "incredible creative energy" that turned her views about Aboriginal people around (chapter 4). Another such source of "incredible creative energy" is the *Lurujarri Heritage Trail*, which nurtured diverse collaborators to successfully oppose the construction of the LNG facility.

Thus far, my theoretical landscape has been mapped out along two pathways: that of transecology as a new theory and method to emphasize biocultural hope and processuality over stasis and a new ethical regime that prioritizes multispecies entanglements over human aspirations. I will now bring these pathways together in a modest reassessment of tourism and heritage as sources for alternative maneuverings within this world of co-becoming.

This concludes my introduction of the theoretical landscape in which the ethnographic analysis of the conflict over Walmadany/James Price Point is embedded. I have proposed as its core facility the notion of transecology. Based on this proposition, my summary will revisit tourism and heritage as alternative "resources" in a transecological project that seeks to highlight an Australian Dreaming after the resources boom. The above has shown that the conventional meanings of tourism and heritage in the Kimberley are twofold. First, they present economic alternatives to the exploration of natural resources. Second, they are alternative "resources" in the sense that they offer means of engaging with the world beyond Modernist constraints. The *Lurujarri Heritage Trail* is exemplary in this regard, as it brings these two meanings of tourism and heritage together. Furthermore, it demonstrates that Indigenous tourism in particular has what I term "hidden qualities" that enable it to bridge the gap between Indigenous and non-Indigenous perceptions of the environment so often evoked in Modernist representations of it (see below).

Initially, tourism was thought of as a mere activity, synonymous with travel. The word "tourist" was first used in 1772, and the word "tourism" in 1811, to describe people who traveled with the intention of seeing different landscapes and experiencing other cultures, languages, and cuisines. Commercial travel for leisure purposes was first associated with the advent of the Industrial Revolution in the United Kingdom—the first European country to promote leisure time to its growing population of industrial workers. Tourism's origins have therefore been located in the origins of industrial development. Gradually, tourism became a more affordable option for the average person.

Meanwhile, in the 1990s, sociologists Scott Lash and John Urry (1994) were already highlighting that innovators in the expansion of the tourism industry, such as Thomas Cook, were as significant to Modernist cultural production as well-recognized figures such as Henry Ford. Researchers became aware of the manifold potentials of tourism beyond that of a mere recreational activity. Today, it is one of the world's most important industries and a key component in the new mobilities paradigm (Sheller and Urry 2006, 2016).

Tourism puts people and places, their heritage and culture, on display. Based on this, an increasingly touristified world continues to create markets where there are none. Qualities of remote regions are readily made available in tourism brochures and advertisements, while Indigenous people and their "exotic" lifestyles become a point of discussion for travelers even before they travel. Tourism therefore remains central to global articulations of people, work, capital, and ideas (Cresswell and Merriman 2011). It is a globe-making activity (Wergin 2017a), a major transformative force (Coleman and Crang 2002), and a preferred policy tool for fostering economic development in remote regions (UNWTO 2009), particularly in Australia.

As an example of its economic potential, in the tourist year ending December 2017, Australia had a total of 8.8 million international arrivals,

6.5 percent more than the year before. International visitors spent $41.3 billion AUS Dollar, constituting 6 percent more than the previous year. In addition, tourism employed 580,200 persons, representing 5 percent of Australia's total workforce. A total of 953,800 international visitors arrived in Western Australia, a figure that went up by 9.1 percent compared to the year before, with an additional 1.4 million interstate and 8.1 million intrastate visitors. Also in WA, $1,526 million AUS Dollar was spent by interstate visitors, $4,026 million AUS Dollar by intrastate overnight visitors, and $2,400 million AUS Dollar by international visitors to the region. Of the international visitors to WA, 630,400 also visited the Kimberley region.

Tourism can therefore be termed a key economic driver for the region, and also for the rest of WA, generating 109,000 jobs and injecting $11.9 billion AUS Dollar into the economy by Gross State Product.[5] Such tourist revenue plays an undeniable role in the development of regional and rural areas, which in turn adds to the diversification of tourism into other niche markets, notably that of Indigenous tourism. Heritage is of equal importance for remote regions and links up with tourism's most common expressions in those, namely eco, sustainable and again Indigenous tourism experiences. In the Kimberley, well-known examples for this are places such as Geeky Gorge, the Horizontal Falls, and the Buccaneer Archipelago. A further example is rock art sites, some of which have been opened up for tourism by their Indigenous custodians and deliberately turned into "sacrificial sites," in the sense that their initial significance within Indigenous Lore and culture has been sacrificed to the tourism industry.

The promotion of heritage experience is thus intrinsically linked to the valorization of Australia's Indigenous heritage and cultural history. The year 1976 marks a scholarly turn in anthropology and the social sciences with regard to the study of such linkages between tourism promotion and heritage production. It is when Dean MacCannell's *The Tourist: A New Theory of the Leisure Class* (1976) described tourism for the first time not as an industry, but as a post-industrial search for truth in "staged authenticities."

In Australia, where the resource industry continues to override the intrinsic values of "living country" and living heritage with promises of economic gain, such a theory of tourism takes a different turn. It spans from the promotion of pristine landscapes to that of iconic objects of alleged everyday use, such as the enormous bush knives that feature in blockbuster movies such as *Crocodile Dundee* (1986). It also includes Baz Luhrmann's *Australia* (2008), which celebrates a Settler-colonial legacy of tough outback heroes and the taming of wilderness through a Western civilizing mission.

Along these lines, the desire to travel to remote regions such as the Kimberley is often stimulated by a wish to experience the extraordinary "natural" heritage of the "last frontier." Heritage sites on related tourist itineraries range from "wilderness" experiences to Aboriginal ceremonial grounds, rock art caves, prison complexes, and the opal mines of Coober Pedy, while other heritage tours to "birthplaces of the resource industry" offer visits to ghost towns along the Dampier coast in WA that showcase resource exploration as the cradle of the modern Australian nation-state.

In other words, Australia is a site of competing "heritages" that broadcast stereotypical narratives of Indigenous "first people" (primitives) versus "pioneers and explorers" (modernists). Heritage and authenticity are therefore deemed complex and contested topics in the Kimberley and beyond. The Australian case also makes evident that they are further means through which to exercise biopower, since many tourism and heritage ventures recycle Settler-colonial myths.

This includes a continued misrepresentation of the violence wreaked on Indigenous people. The pearling legacy of Broome, for example, when remembered in places such as Willie Creek Pearl Farm, is conjoined with the romanticized construction of Multicultural Australia, designed in the bureaucratic minds of strategists located in centers of political power such as Perth (for WA) and Canberra, which leaves a bitter aftertaste when pronouncing the "fair go" that every Australian is allegedly entitled to.

Simultaneously, tourists invest in artworks and ensure the continued existence of markers of Aboriginal presence. Those who attend Aboriginal cultural tours are introduced to bush foods, Aboriginal stories, and sites of historical and cultural significance. Volunteer tourists support Indigenous rangers in maintaining sites, tidying up beaches, and helping to plant and harvest traditional foods. The impact of such collaborative incentives for the promotion of Indigenous tourism experiences reaches significantly further than economic advancement.

Helpful in this regard is the fact that the most successful tourist destinations remain those that can offer experiences "far from the ordinary" (Gibson and Connell 2005, 260), and their success increasingly depends on means by which to overcome the "us versus them" dichotomy. The various accounts I collected from participants in the *Lurujarri Heritage Trail* all express people's desire to engage with Indigenous people, to learn and share, and thus to overcome this conventional dualism.

In other words, while Australian government and industry still seek to turn more of its heritage, Indigenous culture and country into industrialized space, Indigenous tourism experiences offer ways of interrogating the underlying "value discourse" of such decision-making processes on preservation versus exploration of country. More precisely, participants in the *Lurujarri Heritage Trail* (organizers and visitors) share a view of the mainstream economy as "'deflated', not (. . .) their main concern, not their main business" (Muecke and Dibley 2016, 148).

The *Goolarabooloo* avoid mainstreaming their tourism and heritage product according to the needs of late-liberal market economics. Instead, as my ethnographic chapters show, its organizers found ways to reconcile the multiple world(view)s of environmentalists, Indigenous custodians, (international) tourists and travelers, and many others in "living country" and living heritage. This makes for further qualities of Indigenous tourism experiences which, I contend, have thus far been hidden under policies that prioritize economic variables. The section that follows will account for these hidden qualities in detail.

3.2 HIDDEN QUALITIES OF INDIGENOUS TOURISM

Many concerns expressed in environmental movements are symptomatic of wider transformations in the twenty-first century. For example, there is general consensus among scientists, environmentalists, and the wider population that climate change is life-threatening to humans and the earth's fauna and flora as we know it. So far, this issue has been discussed alongside existing categories of global warming, debt, and poverty, overlapping with a critique of the privatization of state welfare, the free-market economy, and Capitalism more generally.

This global movement finds its expression in a growing number of protest movements set up to protect the environment and to reassess the values attached to it. Related emerging collaborations blur conventional boundaries between public and private, us and them, civil and political, Indigenous and non-Indigenous.[6] In this subsection, I explore the potential of Indigenous tourism experiences as further means of fostering such collaboration and therefore as significant contributors to global environmental movements.

Australia is particularly relevant here because its Indigenous tourism product is commonly advertised as what allows participants to "experience the world's oldest living culture" (Vermeersch, Sanders, and Willson 2016, 184; see also Department of Resources, Energy & Tourism 2009, 2011). Meanwhile, there are diverse ways in which this experience is catered for. The National Aboriginal and Torres Strait Islander Tourism Industry Strategy (ATSIC 1997, 3) allows for considerable freedom in this regard when it defines Indigenous tourism as "All forms of participation by Aboriginal and/or Torres Strait Islander people in tourism: as employers; as employees; as investors; as joint venture partners; providing Indigenous cultural products; and providing mainstream tourism products." As Zeppel (2006) points out, such generalizing approaches often lead to the Indigenous cultural experience not only going hand in hand with environmental tourism but being largely overridden by it.[7] Furthermore, Laura Vermeersch, Dale Sanders, and Greg Willson (2016, 186) note that a considerable number of domestic tourists in Australia claim to be knowledgeable about Aboriginal culture and in less need of participation in Indigenous tourism activities (see also Buutljens, Gale, and White 2010). As a result, the majority of tourist travelers interested in Indigenous tourism experiences prioritize learning about nature over learning about Indigeneity (Vermeersch, Sanders, and Willson 2016, 186).

This finding contradicts a prominent debate in social science and humanities research that has identified "Forgetting Aborigines" as a central problem within contemporary Australian society (Healy 2008). With regard to the attractiveness of Indigenous tourism experiences, however, the opposite seems the case—namely "knowing too much," at least about certain aspects of Aboriginal culture and heritage, to be attracted to it. Meanwhile, as will be seen below, there remains a significant part of Aboriginal Lore and custom that is public but still unknown to the wider Australian population, and that becomes accessible with emphasis on its transecological dimension.

Butler and Hinch (2007) define three conventional types of Indigenous tourism products, which are characterized by (1) Indigenous control and theme ("culture controlled"), (2) Indigenous theme but not control ("culture dispossessed"), or (3) Indigenous control but not theme ("diversified Indigenous"). All of the above can be found in the Kimberley. However, its potential as an Indigenous tourism experience lies beyond this conventional understanding that defines it as yet another product designed for the tourism industry. This potential becomes evident when Ingrid J., a consultant to Indigenous tourism businesses throughout the Kimberley, describes her work, the ways in which the types of tourism spelled out by Butler and Hinch play out in the Kimberley, and the challenges this poses:

> I've been in the tourism business for over 20 years, and working in Indigenous tourism in particular . . . for about six or seven years (. . .) I worked as the West Australian coordinator for the Respecting Our Culture-Program, which was run through an organization called Aboriginal Tourism Australia (. . .) [It] was a tourism accreditation program for Indigenous tour operators to assist them to get up to a national standard of tourism accreditation. It also allowed for non-Indigenous operators that had Indigenous content to be able to apply under certain conditions for that (. . .) Part of the aim was to ensure that people that had Indigenous content in their tour—[that] it was authorized by the correct people (. . .) Unfortunately, that folded because of lack of funding (. . .) I thought that was a very good program but (. . .) I suppose it fell out of favour a little bit with some of the supporting organizations (. . .)

> I think one of the things that they didn't understand was that the outcomes didn't quite fit in with . . . mainstream timelines. So they would get a little bit worried that things weren't progressing quickly enough. But bear in mind that we had to go out on country and work with people on their own country. No matter how remote that was. And work through it on *their* time, not *our* time [chuckles]—their timelines, not our timelines. And they might have more important issues going on at the time. There might be family business, or funerals, so they might not want to work with you, and that might go on for six months. . . . And you might have gotten to a certain stage, and then they go, "No, we're not going to do it now" (. . .) And trying to explain that to a funding body, going, "Well, you know, there is two or three operators that I can't work with at the moment for various reasons." And they just look at you like, "Are you kidding me?" And so I think it was hard to justify what they were getting out of it for the money they were spending. You know, in a *whitefella* type way. Whereas slow but steady really is the way to go with developing Indigenous tourism (. . .) It's all very well marketing it, but people have to be able to supply the product that's being marketed. (Transcript 130312_PJ)

Ingrid J. emphasizes that one cannot develop an Indigenous tourism product without the collaboration of Indigenous people. For this to be achieved needs more than good intentions, itineraries, and guidelines to be followed. The goals of such a development scheme must suit the demands of the Indigenous groups that are to be recruited into it. She points to the fact that this takes time and negotiation. Or, in other words, policies and practices cannot be implemented from above but need to be derived from below. Otherwise, the knowledge holders will refuse to participate, because they will have other priorities.

Meanwhile, if organized with not only the consent but also the recognition of the priorities of all parties involved, Indigenous tourism experiences can foster new relationships between people and people, people and place, fauna and flora, that have significant impact on the environment, culture and economic development of a region, as well as the ways in which Indigeneity is perceived within wider Australian society and beyond (see example of the Swiss backpackers, chapter 4). This hidden quality of Indigenous tourism is a political game-changer that has thus far been overlooked because, as Ingrid J. explains, funding bodies set their priorities from a Western point of view and without much understanding of the local circumstances.

The above describes a wider phenomenon that finds its repercussions in development projects around the world, and also with regard to environmental impact assessments of industrialization proposals such as the one for Walmadany/James Price Point. These tend to rely on a distinction between society and nature which sees them as two separable realms and which stands in the way of any acknowledgment of world(view) s and values that do not draw on such dualisms. Indigenous tourism experiences such as the *Lurujarri Heritage Trail* overcome this distinction and stimulate a more ethically balanced approach to people and environments, heritage, and cultural values. Below, Ingrid J., the Indigenous tourism consultant, outlines her involvement in the promotion of related collaborative tourism experiences:

> *Ingrid J.:* I went and worked on the Kimberley Coastal Tourism Project (. . .) a three-year project based around the Kimberley coast and the TO's aspirations for management and involvement in the tourism (. . .)
>
> *Carsten Wergin:* Cruise ships?
>
> *IJ:* Yeah, yeah. (. . .) I ended up working with some of the inland groups as well. And that was mainly because one of the funders (. . .) didn't (. . .) realize that it was actually focused on the Kimberley coast, and they were an operator that worked inland [laughs out loud] (. . .) So they pushed to have some of the inland stuff. But it was a great project. The problem was that it was probably a little bit too bold for one person to try and achieve. So for the amount of money and one person it was too broad, really. But we got through it. And I worked with a lot of the ranger groups, because I was working through the Kimberley Land Council (. . .) in the Land and Sea Unit, through the Traditional Owners (. . .) rather than operators. . . . More through the ranger groups and corporations . . . prescribed body corporates.
>
> *CW:* Have there been any outcomes?
>
> *IJ:* Yeah, there are several outcomes. One of the outcomes was we got as far as a final draft of (. . .) a code of conduct for operators working on the Kimberley coast. We called it something slightly different. We didn't call it guiding principles, but yeah . . . I do understand that since then it has been agreed to by the operators. . . . But it's a voluntary code. . . . How to work on the Kimberley coast (. . .) There was negotiations with the tourism industry and with the Traditional Owners. We did a few site management plans for the hotspots (. . .) So like, Horizontal Falls... Mouth Point, Montgomery Reef, King George Falls. . . . . And then also (. . .) a listing of the hotspots and how the TOs would like to see them managed (. . .)
>
> *CW:* Do all of these big tourist attractions like Horizontal Falls and King George Falls have sensitive places?
>
> *IJ:* Yeah, pretty well (. . .) I think it's natural . . . If something looks fantastic to us (. . .) there is something fantastic in Aboriginal culture that goes along with that. We see the beauty from a white person's sort of point of view. Whereas for them it would be interpreted through a cultural view that would be admiring sort of the same. . . . Other times it might be a burial site. . . . Then there is some Dreamtime story to it . . .
>
> *CW:* Like with the dinosaur footprints (. . .) It might be the story of the Emu Man leaving his footprints, and for the other it's . . . palaeontology . . .

IJ: Yeah, that's right. So everyone thinks it's fantastic.

CW: For various other reasons.

IJ: Depends on what culture you come from, I suppose (. . .)

CW: And are these things integrated into the tourist experience?

IJ: Only if it's known to the tour operator. And then quite often it's only known at this point in time through books and things like that. And that was part of what the project was sort of aiming at: Educating the operators on correct stories and correct protocols. . . . The right people just speaking for those particular areas. A lot of operators at this stage—and it's not necessarily their fault—they might rely on books and things like that that they've researched to get that information.

CW: Couldn't it be an employment opportunity for local Indigenous people?

IJ: Yeah, yeah, potentially, in some of those areas, yeah. Either be it for rangers, for looking after the area, or for engaging in enterprise—as doing guiding and things like that. And that's been part of that project and also part of some of the work I'm doing currently. (Transcript 130312_PJ)

Ingrid J.'s collaborations with the Land and Sea Unit of the Kimberley Land Council, their ranger groups, and also the relevant Traditional Owners rather than tour operators indicate a change of perspective in recent years with regard to the development and implementation of an Indigenous tourism product. Projects as the one described by her seek to right previous wrongs, such as unauthorized tourist access to cultural sites.

Related to this is the continued need of some tourism operators to rely on "books and other things" for relevant information about places they visit, which, as Ingrid J. suggests, "is not necessarily their fault." Meanwhile, there is an obvious reason for Indigenous groups to get involved, if their concerns are heard and their collaboration is ethically balanced and no longer a mere necessity to guarantee the "exotic" experience that their culture is staged to offer. However, as mentioned above, such staging has a considerable history within the tourism market and remains difficult to overcome.

Since the 1970s, when mass tourism began to reach into more remote places, the search for the "exotic" has largely been its *raison d'être*. A global spread of televised documentaries and travel programs, linked to better infrastructures and an increased use of cars and planes, stimulated considerable investment in more remote regions. Various scholars have identified "'exotic' Indigenous peoples with intriguing customs, artifacts, arts and crafts, lifestyles, heritage and histories" as significant attractions in such remote places (Whitford and Ruhanen 2016, 1081; see also Hall and Tucker 2004). This "exotic" appeal corresponds to the initial use of the term "indigenous" to describe "primitive" cultural groups marginalized by the advancement of progress and development (Short 2003).

From the 1980s onward, this marginalization became an issue in government policy documents and reports (Whitford and Ruhanen 2016, 1081). In Australia and elsewhere, Indigenous tourism was seen as a new means through which the central problems facing Indigenous people could be tackled efficiently. The latter account for about 5 percent of the world's population, but about 15 percent of what are considered the world's poor—with their livelihoods characterized by poverty, (racial) discrimination, and conflict (United Nations 2009).

However, Indigenous tourism has not necessarily achieved the desired outcomes, and has therefore been described as a "double-edged sword" (Whitford and Ruhanen 2016, 1082). With this term, Whitford and Ruhanen emphasize the socio-economic disadvantages that Indigenous people are still faced with and that Indigenous tourism products did not help to overcome. Those include the continued exercise of a Settler-colonialist power regime characterized by, as described above, ineffective and misguided government policies, as well as a lack of access to education, health services, and employment (ibid., 1082).

The authors further emphasize that between 1980 and 2000, Indigenous tourism development was still mostly governed by a "neo-liberal political agenda" (ibid., 1084). Among other things, this meant that the involvement of Indigenous groups in the design process of such programs was marginal to non-existent, even though practitioners like Ingrid J. highlight its necessity.

Government involvement was equally limited in Australia until, in 1997, the National Aboriginal and Torres Strait Islander Tourism Strategy mentioned above was published. The strategy was intended to push toward the better alignment of Indigenous tourism with the mainstream tourism industry (Whitford and Ruhanen 2016, 1084; see also Schmiechen 2006; Whitford, Bell, and Watkins 2001; and above). As a result, research into Indigenous tourism began to focus more on issues related to the continued impact of Settler-colonialism, the commodification and preservation of nature and culture, and identity formation and authenticity.[8] Furthermore, these projects raised concerns about the empowerment of Indigenous people through tourism and related questions of Native Title, land tenure, and access to necessary funds, and these remain of central importance.[9] In Australia, this was linked to issues of sustainability, with emphasis on the qualities of (Indigenous) control, ownership, and management rights for the procurement of cultural preservation and authenticity in the light of Native Title, the land rights movement and the impact of globalization more generally (Altman 1989; Altman and Finlayson 1992; Gale et al. 1988). Here, Whitford and Ruhanen (2016) point out that, "The common thread [. . .] was the recognition of the need to adopt and implement a bottom-up approach to Indigenous tourism development that provided legitimate Indigenous community and/or individual participation in planning, management and ownership of Indigenous tourism products and services" (Whitford and Ruhanen 2016, 1086). Researchers became increasingly aware of numerous "wicked problems" within this sustainability/Indigenous tourism nexus (ibid., 1086; see also Head 2008; Scherrer and Doohan 2014). Though much of the research conducted in Australia sought for ways to integrate sustainability measures into Indigenous tourism products, this was barely achieved, because Western policy and Indigenous practice continued to operate out of two different sets of values and world(view)s.

Meanwhile, there is general criticism that federal and state governments have underplayed or ignored the risk to tourism that is posed by climate change, greenhouse gas emissions, and related needs to improve sustainability measures. This is linked to an overemphasis on economic growth instead of the preservation of heritage sites that function as lucrative tourism products. Broome is a site where this conflict becomes apparent on a daily basis, as it is both the main service center for the broader Kimberley region and a key tourist destination in WA.

Up until 2012, the Broome tourism economy had averaged 3.1 percent growth per year over the previous four years. Meanwhile, there was significant fluctuation in the number of visitors to Broome between 2001 and 2012, ranging from a low of 200,883 visitors in 2010 to a high of 338,203 visitors in 2003. However, between 2011 and 2012 (Year ending June 2012), Broome saw an increase of 20 percent in domestic overnight visitation (AEC Group 2012). The town therefore remains reliant on tourism, with retail trade, transport, accommodation and food services contributing roughly 30 percent to GRP and employment in the area.

The tourism sector furthermore remains highly seasonal, which means that many businesses operate at very high levels during peak season (April to September), while their business falls significantly during the low season (October to March). The wet/low season experience in Broome has been characterized in tourism assessment reports as a "not-open-for-business attitude" that negatively impacts visitors' perceptions of the town (Haeberlin Consulting 2014). This stands in stark contrast to the dry/peak season and its "buzz and vibe in town that needs to be carried on to other times of the year" (ibid.).

Indigenous tourism is generally more sought after by international travelers, although domestic markets are usually larger, with Australia being a case in point (Abascal, Fluker, and Jiang 2015; Ryan and Huyton 2002). However, between 2006 and 2010, Australia experienced an 18.7

percent average annual decline in domestic overnight Indigenous tourism visitors and a 4.9 percent average annual decline in international Indigenous tourism visitors (ibid., 1087; see also Tourism Research Australia 2011). Ruhanen, Whitford, and McLennan (2015a, 2015b) confirmed these lower levels of demand and of interest in the Indigenous tourism product for 2012. Meanwhile, with regard to the ethnographic data discussed here, Broome and the Dampier Peninsula present particular cases in point, as their tourist numbers stayed the same and even increased over the same period.

This difference is due to the fact that Broome and the Kimberley have qualities that set them apart from other regions in Australia. Those include its distinct Creole culture that has mixed Aboriginal, Asian, and European influences since the town's original settlement in the late nineteenth century. Furthermore, the qualities of the local environment have provided the readily marketable appeal of pristine wilderness mixed with white sandy beaches and spectacular sunsets. Indigenous tourism products are particularly suitable for bringing together the cultural and environmental qualities, not least because Indigenous Australians continue to supply the exotic appeal of "primitive" in such tourism ventures, offering a point of entry into a relationship with nature that is "unspoiled by development" and that the "civilized" have left behind. This has not only occurred in Australia, but has been linked to increased displays of "professional primitivism" (Bastin 2003), and new emerging markets such as "reconciliation tourism" (Higgins-Desbiolles 2003) throughout the world.

In sum, Indigenous tourism research has grown exponentially over the last thirty-five years (Whitford and Ruhanen 2016, 1089), while Indigenous tourism has been termed a vehicle for re-indigenization and part of a broader contemporary Indigenous renaissance. Ameliorating the relationships between Indigenous and non-Indigenous Australians remains a common goal in such Indigenous tourism incentives (Weaver 2016, 32). However, there also remains the need to depart from the use of Indigenous tourism as an additional tool for policy implementations (Whitford and Ruhanen 2016, 1090), because, as seen above, it is not policies that will drive the development of Indigenous tourism products, but that which the market has to offer—that which is already in place, namely people, environments and their heritage values.

Because of this, the particular components of an Indigenous tourism experience like the "living country" that is traversed during the *Lurujarri Heritage Trail* need considerably more attention from the start. The recognition of "living country" fosters in participants perceptions of the environment that go beyond the nature/culture divide. This alone demands that contemporary research and policy offer perspectives on the potentials of Indigenous tourism experiences that go beyond prevailing late-liberal market interests such as closing the economic gap between Indigenous and non-Indigenous people (Wergin 2018b).

Along these lines, researchers have emphasized that participation in Indigenous tourism experiences allows Australian domestic tourists to take time and reflect on their "home, Aboriginality, Aboriginal issues and themselves" (Gallifort 2010, 242). David Weaver (2016), in his discussion of what he terms "nonconventional indigenous tourism (NIP)," explains that a tourism product that carries the title "indigenous" is commonly associated with stakeholders who hold such status under international law. These "conventional indigenous people" (CIPs) (Weaver 2016) number approximately 370 million across about 70 countries and hold at least 5,000 distinct cultures worldwide (UNPFII 2015). However, Weaver proposes "a more liberal definition of indigeneity and subsequently an expanded product and knowledge domain of 'nonconventional' indigenous tourism" (Weaver 2016, 29). He sees a need for such a reassessment not least because of the aforementioned global spread of "re-indigenization," or, "the rediscovery, reassertion, and reclamation of an indigenous culture, or elements thereof, by members claiming that culture [. . .] a partial pull-back from modernity toward tradition (real or imagined) following a period of cultural alienation associated with hybridization or assimilation" (ibid., 32; see also Attanapola and Lund 2013).

Weaver's analysis ties in with the increase in environmental movements and the motivations of people who participate in them. As stated above, Indigeneity has in this way become an indicator of a generally more "earthbound" approach to the world (Latour 2017). This also holds true for non-Indigenous people who seek to retrieve entangled connections to the environment. Indigenous tourism experience can equally be termed "nonconventional" in the sense that people who might previously have defined themselves as non-Indigenous can become stakeholders in it, if they adhere to the transecological values it entails. Here it is worth noting that since its inauguration in 1987, the *Lurujarri Heritage Trail* has always relied on the active participation of non-Indigenous people in it.

Despite such "nonconventional" theoretical approaches, it is necessary to remember that the late-liberal mainstream still overrides many of the related incentives. More often than not, resource explorations come with the exploitation of people, land, and sea. Here, tourism and service sector industries, particularly in remote regions in Australia, suffer significantly as a result of mining developments. For one, such developments foster a lack of available workforce in these sectors because of their higher pay rates. In addition, the crowding out of flights by FIFO workers leaves insufficient seats available for leisure tourists. As mentioned above, industrialization projects also have a largely negative impact on local communities in general, as they result in, for example, increased male dominance and a related change of atmosphere in restaurants, bars and pubs.[10]

In contrast to this, the lesson taught to participants in the *Lurujarri Heritage Trail* through the concept of "living country" is that of an entangled relationship between people and place. Participants share an acknowledgment of the land as inseparable from society to the point where some seek to preserve the local environment themselves, counteracting the attempts of economic and political elites to dominate decision-making processes about it. Learning to engage with the environment as "living country" thus generates sound opposition to the approach advocated by the resource industry that puts into action a FIFO workforce to "cultivate" the land.

An outside *force* can be destabilized through Indigenous tourism experiences like the *Lurujarri Heritage Trail*, which offer means to re-Indigenize people and place in experience. The growing body of literature at the intersections of critical geography, heritage studies, and anthropology, some of which I have presented above, adds to this to further the immanent critique of both dualist thinking and a related dominance of Western worldviews over others. Many such theorists acknowledge instead the co-becoming of people with the environment as a means of evading essentializing discourses (see above).

My ethnographic material about the *Lurujarri Heritage Trail* and the lived experience it offers demonstrates how such theory and practice complement each other. The qualities of a transecological engagement with people and place give new impetus to the wider social sciences and policy research. During the *Lurujarri Heritage Trail*, "living country" is experienced not as a stable substance with clearly defined boundaries ready to be explored, but as that which emerges from collaboration between the human and the more-than-human. The Environmental Humanities draw on such perceptions of the environment to offer new scientific concepts and/or re-appropriate old ones.

Karen Barad provides an example of this when she redefines the main goal of differentiating as that of "making connections and commitments" (Barad 2012, 47). Again, it is important to note that a critique of dualisms does not imply a rejection of the usefulness of binarisms. Differentiating remains an important part of critical thinking. What is relevant here is that to make a difference, and with that to create binaries, remains just that: a means to structure our thinking about complex phenomena that cannot be expressed in binarisms alone.

Related theoretical notions that I draw on in my analysis include Barad's agential realism, which is a further means of describing fruitful alliances within and outside academia that reveal flows of material agencies across regions, environments, animal and human bodies. With regard to the qualities of an "unconventional" Indigenous tourism experience like the *Lurujarri Heritage Trail*, the point to be made here is that such practice adds to the discussion concepts such as "living country" and living heritage, further broadening the theoretical landscape.

I conclude this chapter with relevant quantitative data on the economic impacts of mining and tourism in Australia, which in strictly Modernist terms provides further evidence for why the industrialization proposal was misguided by an opportunity dreaming that failed to recognize the values already in place. "Mining versus Tourism" was a central dualism used by proponents of the LNG facility to argue for its construction, or, as the then Broome Shire president put it, "tourism won't save Broome." The data presented below are relevant in this regard, as it shows that while mining has been a key driver of Australia's national economy, it has not played a significant role in the Kimberley, let alone in Broome. Instead, tourism has established itself as a thriving economy both in the Kimberley and on a national level, with twice as many employees as the mining sector. Relevant quantitative data are available to counteract the alleged benefits of the mining boom for the Kimberley.

3.3 FURTHER FACTS ON "MINING VS. TOURISM" IN AUSTRALIA

Conflicts between diverse value regimes and related world(view)s have a long history of playing out in the Kimberley. With colonial settlement, the region became "crown land." Settlers erected cattle stations and fences. Broome turned into the largest export harbour for pearl shell in the world. When plastic replaced mother-of-pearl as the material used for the production of buttons, Broome had to reinvent itself; it became an access point for adventurous tourists and travelers to the remote Dampier Peninsula and beyond, "the last frontier," and the "gateway to the Kimberley."

In 2006, Broome saw yet another attempt of reinvention: a $45 billion AUS Dollar LNG proposal that would turn it from tourist into mining town. Numerous NGOs, most notably EK, and diverse members of Broome's civil society collaborated to raise awareness of the serious environmental, social, and cultural concerns related to this proposal. Their fight continues to this day, in conflicts over fracking, uranium mining, and arguments for the Kimberley to become a possible "Food bowl of the North." The latter has been enhanced via significant Chinese investment in its agro-economics, which has seen the first attempts to cultivate rice in this semi-arid region take place.

In order to put the potential of such attempts at reinvention into a quantifiable perspective, this subchapter summarizes official government data available on the contributions of mining and tourism to the Australian economy.[11] To begin with some facts particularly relevant for this ethnography, in 2016 there was no documented employment by the mining industry within the Broome Local Government Area (LGA), resulting in a gross value add per mining employee to the LGA of $0 AUS Dollar.

Meanwhile, in 2017, tourism in the Kimberley estimated 34,600 (49%) international, 89,300 (39%) interstate, and 144,900 (15%) intrastate visitors to the Shire of Broome alone, with the total number of overnight visitors to Australia's northwest spending $696 million AUS Dollar. Earnings (as of March 2017) for the whole of WA were at $1,526 million from interstate, $4,026 million from intrastate, and $2,400 million from international visitors.

Furthermore, on a regional level, in 2016–2017 the $356 million AUS Dollar total share of WA mining production from the Kimberley region represented only 0.34 percent of the state's total, with Broome contributing $4 million, compared to a contribution of $63 billion from the Pilbara region. This is despite the fact that since 2006–2007, investment in mining had grown significantly in the Kimberley with the increase in mining and gas projects off the coast. While the proposal for an LNG facility at Walmadany/James Price Point was officially abandoned in 2013, a floating LNG platform in the Browse Basin was still expected to begin drilling in 2017.

On a state level, in 2017 the volume of LNG produced in WA increased by 38 percent, reaching a record 32.7 million tonnes. LNG has been termed WA's most valuable petroleum product, accounting for 14 percent of the State's total mineral and petroleum sales in 2017. In the same year, the value of LNG sales value rose from $10.6 billion to $14.9 billion AUS Dollar. Meanwhile, the value of projects under construction or in the committed stage of development was an estimated $54 billion, down from $96 billion, which was also due to the abandonment of the Walmadany/James Price Point project.

WA's mining industry directly employed an average of 102,258 people during 2015–2016. The state's petroleum industry, which LNG is part of, employed an average of only 1380 people in the same time period, which was down 33 percent from an average of 2058 in 2014–2015. On a national level, the mining industry employed approximately 230,000 people in 2016, with more employed by related industries, and accounted for 7.4 percent of the gross domestic product (GDP).

For the same year, the national number of tourism employees was more than double at 580,200, which is 5 percent of the total Australian workforce. Earnings from international visitors mounted to $40.6 billion at December 2016 and $41.3 billion at December 2017 (up 6%). Here, Australia's mineral exports (excluding petroleum products) amounted to over $151 billion, which was almost 58 percent of all export merchandise and 46 percent of all exported goods and services.

In sum, while the resources industry makes a significantly greater contribution to Australia's GDP, the mining sector employs significantly fewer people, namely less than half of those working in the tourism sector. Meanwhile, in Broome and the Kimberley, tourism not only employs significantly more people but also makes twice the contribution to the local economy, with $696 million spent by overnight visitors to Australia's Northwest compared to the $356 million generated by mining in the Kimberley.

The data furthermore highlights that the Shire of Broome is a preferred destination particularly for intrastate and international visitors to WA: 39 percent of intrastate and 49 percent of international visitors who would visit WA would (also) visit the shire of Broome.

NOTES

1. https://www.dss.gov.au/sites/default/files/documents/03_2017/multicultural_policy_2017.pdf (Date retrieved: March 16, 2018).

2. A so-called "skin system" subdivides Aboriginal society into named categories that, among other things, traditionally prevented some people from marrying certain others. There remain rules that are still adhered to, such as "avoidance relationships" like that between a mother-in-law and a son-in-law. For further reference in regards to the *Yawuru* people of the Broome area, see https://www.yawuru.com/our-culture/family/ (Accessed November 15, 2018).

3. For one, the nature/culture dualism asserts that nature has no agency other than that attributed to it by human intelligence. Furthermore, this dualism undermines a critical engagement with its very origins in Modernist thought.

4. A concrete example of this is the resistance that proponents of the industrialization project for Walmadany/James Price Point were confronted with, and that created such significant uncertainty among investors that some withdrew their commitment even before any final decision was made.

5. Most of the 2017 international visitors to Australia originated from the United Kingdom (16%), then Malaysia (12%), Singapore (11%), the USA (8%), New Zealand (8%), China (5%), Germany (4%), Indonesia (3%), Japan (3%), and Hong Kong (2%). Other Asia makes up 20 percent, and Other Countries 5 percent. For further references, visit: http://www.tourism.australia.com/en/markets-and-research/tourism-statistics.html (Accessed March 13, 2018). See also: http://www.tourism.wa.gov.au/Research-Reports/Latest_Visitor_Facts_and_Figures/Pages/Economic-Contribution-of-Tourism-to-WA.aspx (Accessed November 13, 2021).

6. One of those conflicts, which involved diverse Indigenous and non-Indigenous groups, government and industrial stakeholders, and became a heated subject in international mainstream and social media during this writing process, was the construction of the Dakota access pipeline, a 1,886 kilometers underground oil pipeline in the United States that became operational on 01 June 2017. For further information concerning these protests, see the widespread online discussions at #NODAPL.

7. Higgins-Desbiolles (2009) in a case study of Camp Coorong, South Australia, offers a critical perspective on this entanglement between Indigenous and nature-based tourism experiences and their impact on the ecological consciousness of participants.

8. See, for example, Adams 1997; Altman and Finlayson 1992; Hinch, Delamere, and Reid 1993; Dragovich 1993; Blundell 1993; Butler and Hinch 2007; Loverseed 1998; Notzke 1999; Steinberg 1994; White, Williams, and Hood 1998; Williams and Stewart, 1997.

9. See, for example, Brown 1999; Hemming 1994; Hohl and Tisdell 1995; King 1994; King and Stewart 1996; Loverseed 1998; Ross 1991; Schuler, Aberdeen, and Dyer 1999; Wall 1996; Wells 1996; Wesche 1993; White 1993; Zeppel 1998.

10. In all 6,000–8,000 workers who were to build the facility at Walmadany/James Price Point over the course of five to ten years would clearly have had a significant impact on not only restaurants, shops or at Cable Beach, but also the social fabric and general iconic status of the tourist town of Broome.

11. The appendix provides a detailed overview of these facts and figures relevant to the wider discussion of the possible benefits and pitfalls of mining and tourism. It presents comparative data on regional, state and national levels.

Chapter 4

The Four Pillars of Settler-Colonialism

Australian ethnography and land rights debates have shown that in Aboriginal societies kinship is inseparable from economy, polity, and religion (Myers 1987; Lea 2012). Likewise, the collaborations that are at the heart of my ethnographic analyses tell of a complex entanglement of people and land, and Indigenous world(view)s that in parts of Australia still prevail despite the cultural violence introduced through what I call the four pillars of Settler-colonialism, which are: dualisms, linear time, ownership, and naming.

In this chapter, I introduce these four pillars and outline how my ethnography provides evidence for world(view)s that have not fallen prey to them. I show that kinship ties, and with them ontological difference, cannot be overridden solely by economic opportunism. Since the famous Mabo decision of 1991—the precursor to Australian Native Title legislation—specific, vested families have been responsible for, and responsive to, particular landscapes and their heritage, through Aboriginal cosmologies such as *Bugarrigarra* and an attachment to the land that goes significantly beyond monetary interests.

Meanwhile, the fear of an inevitable loss of Aboriginal culture to Settler-colonialism has motivated a significant amount of research(ers), for a significant amount of time. The German anthropologist Helmut Petri (1907–1986) used the term "contact situation" to describe the context in which he found Indigenous communities and himself when he first conducted fieldwork in North-West Australia in 1938 (Petri 1987, 619). He expressed his concern about the impact of Settler-colonialism and argued that the contact situation set in motion an acculturation process that would ultimately see Indigenous cultures merge with and disappear into Western civilization.

These consequences of the "contact situation" continue to nurture a general critique of Settler-colonialism, ninety years after Petri first visited the Kimberley. In various domains such as Indigenous rock art, the history of the Stolen Generation, and Native Title determinations, Settler-colonialism continues to override Indigenous Lore and culture with a prevailing narrative of Western law, education, and mainstream economics, legitimized by the allegedly shared values of Multicultural Australia.

Challenges faced by Indigenous people who seek to set up a tourism business provide concrete examples of the ways in which Settler-colonialism plays out today. An anonymous employee of the Kimberley Training Institute, an adult education center, provided a vivid account of this, which I reproduce in the following interview excerpt:

> *Interviewee:* I was offered a job up here to set up a course called Retail Tourism for an organization called Hedland College and I came to Broome for a three-month contract. It was a 12-week course. I came from industry, I didn't really have that much of a training background.
>
> *Carsten Wergin:* When was that?
>
> *I:* In 1995. And during that course I had a lot of long-term unemployed people and also about three or four Aboriginal people on the course. And they all completed to the end and most of them got jobs (. . .) And no one's ever got Aboriginal people through this course and made it interesting and got them involved and got them working in the industry before. So then they extended my contract to 12 months and we started going camping and fishing (. . .) We had the training manual that was basically training people to be tour guides. The companies in Broome that were all owned by non-Aboriginal people wanted those people to be guides to work on the buses. So they wanted to drive them to Beagle Bay [Aboriginal community and former mission station on the Dampier Peninsula, *the author*], an Aboriginal person jumped on, got paid, talked to the tourists and got off again.
>
> *CW:* Good idea.
>
> *I:* Yeah, it's a good idea, but (. . .) Aboriginal people didn't want to be there at 12:00 when the bus comes. Because it might be good for fishing today, or it might be good for camping today, or it might be good for hunting today. . . . One of their family might be sick. . . . There might be so many things happening in the communities (. . .) What [Aboriginal people] wanted was the skills and knowledge to do the business. They didn't really care whether they had a certificate or not. So we adapted the model so that we holistically assessed people in a way that they could develop their own training around their own idea. The first thing I asked people was, "Who's ever been a tourist?" and "Who's been on a tour?" And nobody had been on a tour. And then I said, "Who's ever been on holidays?" And people went, "I have a sister in Panna-

> wonica near Port Hedland. I stayed there once." "What did you do?" "Well, I sat in the house and watched videos all day." "Did you go on a tour?" "No." "Has anyone ever been on a fishing charter?" "No." "Has anyone ever been on the camels in Broome?" "No."
>
> So what we did was develop this model around, firstly, giving people the experience of knowing what it was like to be a tourist. Because how could you know what people would want if they didn't understand what a tour actually was? We developed a model that ran for twelve months and it was a Certificate 2 in Tourism and it involved forty weeks' training on-site in their community, and then six block releases where they came into Broome for a week. And during those weeks we would do things like go on a tour, look at the hotels, go out to dinner at a restaurant. "Who's ever been to dinner at a restaurant?" Nobody, you know? And during those things we interacted with other tourists. And we started to talk to the other tourists, especially the international tourists who were very interested in Aboriginal culture, in Aboriginal people, and for a lot of non-Aboriginal people it was the first Aboriginal person they'd ever met. . . . For a lot! (Transcript 121122_KTI)

From this account, the contact situation no longer appears as a matter of concern. Quite the contrary: Aboriginal people interested in the development of their own tourism products lack the contact with tourism experiences and tourists that is necessary to get a sense of what the industry seeks. As will be seen below, the interest of tourists and travelers in Aboriginal experiences can support the preservation of Indigenous culture rather than overriding it, if these experiences are designed using collaborative means as the *Lurujarri Heritage Trail* does.

Taking an insider perspective allowed the trainer to restructure his course appropriately. He focused on what Indigenous participants found useful, not on what government policy assumed to be valuable for them. People became motivated because the trainer allowed their motivation to develop from within. Previously, the same people had been told what to do in order to succeed in Settler-colonialist society. What they had been waiting for was this possibility of making an active contribution rather than being solely the recipients of benefits.

Values that are fundamental to the imaginary of a Multicultural Australia had been, not shared, but indoctrinated from above, albeit without the anticipated results because Aboriginal people refused to participate. The *Lurujarri Heritage Trail* is emblematic of alternative means by which Indigenous groups make their voices heard rather than relying on government funding and policy. It opens up collaborative spaces for onto-epistemic partnerships with non-Indigenous groups.

Emblematic of those partnerships "outside the mainstream," in the Kimberley region and elsewhere, are the ways in which Indigenous knowledge is presented and preserved in transcultural encounters. The anonymous employee of the Kimberley Training Institute continues his account with another graphic example of the necessity for these encounters to overcome superficial impressions that lead to stereotypes and false generalizations:

> I used to take backpackers as practice tourists (. . .) Because all the Aboriginal people, from the bush especially, when they come to town, they don't necessarily interact with non-Aboriginal people. They only interact with their family and extended family and they go to network within that community. So they're not exposed, generally speaking, to the wider non-Aboriginal community. They see it on TV. But as far as sitting and having a yarn like this, people are a bit funny. . . . So I started to bring those practice tourists. And I got these Swiss girls. . . .We picked them up in Broome, we drove to Beagle Bay (. . .) And I said to my students, "I want you to take them on a small tour (. . .) You don't have to write it all down. You just tell them what you know about the country, and about the history, and where you were born."
>
> And I found, once people started telling their story, they were much more comfortable. So we were all in this big [Toyota] troop carrier. And there was the two girls and me, I'm driving, and there was eight or nine people. There was a lot of grandparents, older people who wanted to learn the skills. And at the end of the day (. . .) I filmed this girl, and I said to her, "What did you learn today?" And she looked me in the eye and she said, "Today, I learned that not all Aboriginal people are drunks." And my students went, "What?" And I said, "Why do you say that?" And she goes, "When I left Perth I went to Geraldton and all I saw was drunk Aboriginal people in the street. I went to Port Hedland, I went to Canarvon, and all I saw was drunk Aboriginal people in the street. When I stopped in Port Hedland, everywhere I went, even in Broome." (. . .) Well, my mob up the Peninsula . . . most of them don't even drink. They were shocked! And she was just matter of fact. So then it started a discussion around alcohol, family ties, all those things, and I saw the separation between the two. (Transcript 121122_KTI)

This account highlights the relevance of the contact situation in helping to overcome prevailing stereotypes such as the widespread misrepresentation of Aboriginal people as drug and alcohol abusers. The case furthermore hints at the inherent potential of Indigenous tourism experiences to foster new, unconventional partnerships beyond the dualism between Indigenous and non-Indigenous that is multiplied in the concept of a Multicultural Australia. Such dualism has negative repercussions in the form of environmental mismanagement, cultural degradation, and everything else.

Our current era, which some have termed the Anthropocene (Crutzen and Stoermer 2000), Capitalocene (Moore 2015), or Chthulucene

(Haraway 2015), sees dualisms of matter and life, inorganic and organic, passive object and active subject crumble. Such dualisms have failed as adequate tools for understanding world-making activities beyond the teleological-minded futurism that is dear to the Modernist project (DeLanda 2000). A new agenda that proffers the ontological turn seeks to depart from this anthropocentric position through a shared commitment to the project of decolonization. Relevant to the reinterpretation of the notion of ontology and how it plays out in world-making activities is that emphasis is placed on the ways in which people define themselves through concrete things, concepts and feelings of belonging, rather than abstract concepts of identity or meaning (Muecke 2012).

Such emphasis on experience opens up the concept of ontology for research(ers) into interrelated world(view)s without assigning more importance to one world(view) than another. While this has been heralded as more egalitarian than dualist thought, considerable criticism of the ontological turn has also been put forward. Zoe Todd (2016), for instance, agrees that the ontological turn can help to overcome dualisms such as nature/culture and human/non-human. But her critique is that representatives of this theory fail to take into account that many such dualisms have been rooted in Modernist thought from the outset. In other words, "the Moderns," as Latour (2013a) calls its proponents, advocate the ontological turn as a legitimate critique of themselves. However, their point of critique—false dualisms and related misrepresentations—was never the main concern of most people in the so-called Global South, up until the North inflicted it on them through slavery, Settler-colonialism, and contemporary late-liberal market economics.

Consequently, for Todd, more emphasis needs to be put on the fact that the very problem that the ontological turn seeks to tackle—the question of how to reconcile multiple world(view)s—is a predominantly Western one. As such, the ontological turn can only adequately reach its goal if it also tackles the larger decolonization project. For this to work, advocates of the ontological turn need to differentiate their criticism from Indigenous knowledge and practice. The latter have never been assumed to be modern, but continue to be pushed in that direction, as the previous examples from the training of Aboriginal people in tourism enterprises show. In other words, as long as a critique of dualisms remains the central concern for the academic world at large, the ontological turn will continue the very Modernist project it seeks to overcome.

To take such criticism seriously meant that I needed to be attentive to the possible need to decolonialize the theory that my ethnographic writing draws upon. I consider the ontological turn valuable for my engagement in the diverse collaborations that emerged in the course of the Walmadany/James Price Point conflict, because to emphasize the coexistence of diverse world(view)s challenges research(ers) to imagine possibilities and accommodate them in their writing. This challenge entails more than the comparison of ontologies. It demands "a technology of description designed in the optimist (non-sceptical) hope of making the otherwise visible" (Holbraad, Pedersen, and Viveiros de Castro 2014, n.p.).

Particularly useful for my ethnographic enquiry is Aileen Moreton-Robinson's (2003) description of this "otherwise" in regard to Indigenous peoples' sense of place. Herself an Indigenous scholar from the *Quandamooka* First Nation (Queensland, Australia), she shows in her work how Indigenous perceptions of the environment are configured differently from those of Settler-colonialists. Indigenous people, including Indigenous Australians, do not assert ownership and moral authority over the land because their sovereignty is ontologically tied to it (Moreton-Robinson 2003, 37). Instead, they consider themselves dependent on "country"—that is, as Australian anthropologist William Edward Hanley Stanner defines it, "land already related to people" (Stanner 1965, 14).

In such entanglements, a "healthy country" is crucial for the health of its Indigenous custodians, and vice versa (Rose 1996). Anthropologist Deborah Bird Rose and others emphasize these "nourishing" qualities of the land that are both physical and spiritual. Her notion of "nourishing terrains" (ibid.) is close to what the *Goolarabooloo* term "living country," and one to which I turn further on in more detail.

What is relevant to the present discussion of ontology as a means to make the "otherwise" visible is that the entanglement with country automatically leaves dualist understandings of people and land, nature and culture, behind because of its emphasis on interdependence. Meanwhile, in the case of Walmadany/James Price Point, the resources industry provides for a particularly illustrative example of how human impact on the environment ignores such entanglements in the pursuit of economic gain and thus (re)introduces dualisms.

Like no other industry in Australia, the resources industry qualifies as both symbol and motor of Settler-colonialism in a search for economic opportunities that are legitimized by the Modernist project. Fundamental to this is a perception of the environment that is based on the nature/culture divide and an image of humans as masters of the environment. This is ontologically different from the Indigenous understanding of country because, in the terms outlined above, the Modernist project rejects the idea that the land is nourishing in itself and instead advocates the need for its cultivation in order to become nourishing.

In the Kimberley, cultivation began with agriculture and pastoralism, but today digs significantly deeper by means that include the use of the controversial technology of fracking. Meanwhile, the Indigenous world(view) is rooted in a *care-full* interrelationship of people and land

(Wergin with Mulligan, 2021). *Bugarrigarra*, the aforementioned West Kimberley term for Aboriginal ancestral Lore and culture, adds to this acknowledgment of the entanglement of people and place a temporal dimension, as it extends equally into the past and into the future, while it is also ever-present in the here and now through ceremony.

Kimberley Indigenous Lore and culture stemming from ancestral beings are linked to particular places and are passed on in ceremonies and customs, stories, and experiences. Lore and culture have been enacted, as Indigenous collaborators explained to me, "since day one." Previous scholarship has approached this interrelationship of people and place, past and present, human and other-than-human, and its inscription in Indigenous Lore and culture, through the notion of "totemism":

> a mystical connection, expressed by symbolic devices and maintained by rules, between living persons, whether as individuals or as groups or as stocks, and other existents—their "totems"—within an *ontology of life* that in Aboriginal understanding depends for order and continuity on maintaining the identities and associations which exemplify the connection. (Stanner 1979, 127–28, *emphasis added*)

This quote from anthropologist W. E. H. Stanner highlights how the study of totemism and other types of "myth-consciousness" relate to the contemporary concern with ontological differences (Kolig 2000, 10). It argues that Indigenous perceptions of the environment are tied to the past and the present equally not only through myths and ceremony but also through daily activities like fishing. What Stanner and others explained as "totemism" provided a means for Indigenous people to inscribe their ontologies in an environment in which the past and the present are interlinked (Swain 1993).

The Settler-colonialist project has done much to override this world(view). First, it introduced temporality as a linear perception of time, with obvious impact on Indigenous societies. According to Modernist logic, the land could not be nourishing in itself. It needed cultivation. In other words, the land had no value without "improvement." The implications of this were severe when viewed through an Indigenous ontological lens that advocated an entanglement of people and land: if the land had no value in itself, it equally meant that "it was not really worth much being an Aboriginal in those days [either]."[1]

Before Settler-colonialism, the value of an Aboriginal person was unquestionable, since people were entangled with the land and as such intrinsically valuable. Afterwards, to term oneself Indigenous became backward-looking, somewhat frozen in time, without the necessary will to strive for a prosperous future. In 1992, the famous Mabo decision partly recognized this: for the first time in history, the Supreme Court of

Australia acknowledged an Indigenous person and his extended family as the owners of land prior to colonization. Some interpreted this decision as a first step toward reconciliation, powered by the return of the land to its original inhabitants. However, Native Title remains based on the recognition of ownership. Its foundations therefore remain within the constraints of a Settler-colonialist view that sees people hierarchically positioned above the land (as its potential owners) and not entangled with it (Wergin 2021).

People and land remain separate entities before the law that makes Native Title possible in the first place. Native Title can therefore only continue to advocate a dualist understanding of people and land, because it can only determine the people who own the land, but not vice versa.[2] In regard to the impact of Settler-colonialism on Indigenous people, ownership thus becomes a third transformative power, together with dualisms and the introduction of a linear perception of time, along which lines Settler-colonialism and with it the cultural violence of the contact situation between colonizer and colonized unfolded.

A striking experience for me in this regard was when I asked an Indigenous person about the contemporary importance of kinship rules for her and in her skin group. She replied that many of her peers would determine their kin not in relation to their skin but in relation to "where the money is."[3] In other words, many would ascribe value to a particular side of their Aboriginal heritage based on mainstream economic interests, rather than intrinsic qualities of their skin that pertain their entanglement with others and a particular part of the land. This is where the interests of the resource industry come into play.

Australia has a long history of mining, and its extractive industries still seek to expand. However, with more areas in remote Australia coming under Native Title law, government and industry have had to look for new ways to access this land. Large compensation packages for people and communities are one of the preferred means. The values of an ontologically different world(view) therefore continue to be overridden by the use of the same transformative powers that were introduced by Settler-colonialism: dualisms, linear time, and ownership still prevail today, which is emblematic also for the conflict that unfolded over the industrialization of Walmadany/James Price Point.

Aboriginal people, knowledge, culture and kinship are traditionally entangled with the land. They are embodied through tracks, designs, obligations, and rights that tie bodies to places and practices (Attwood 1996; Merlan 1996, 1998). The question I came to ask myself was, how to compensate for the loss of such entanglement? The answer to this question is linked to the fourth and final pillar of Settler-colonialism: the introduction of a Western language that led Indigenous groups and individuals to expect "a tentative ontological status" from naming (Butler 1997, 25).

In Australia, missionaries played a significant role in this as they taught Indigenous Australians Church songs in new languages, gave them Christian names, and restructured daily routines. Many influential Indigenous Australian politicians, artists and academics have been through such schooling. National figures include Marcia Langton, Archie Roach and Noel Pearson; some have close ties to the Kimberley, like Stephen Pigram, Patrick and Mick Dodson, and Peter Yu. Boarding schools run by the Church are where these influential thinkers learnt to "talk the talk"—to understand the rules of the power-knowledge game of the West versus the rest—which has put them in a privileged position to advocate for reconciliation. In other words, the colonized talk back.

Their careers might have begun at a time when the Modern assimilationist policy that was advocated in schools taught them that "it was not worth much being an Aboriginal." However, in the age of Native Title, climate change, and environmental degradation, Aboriginal knowledge and culture add an important component to contemporary politics and policy. Luckily, none of the above Indigenous experts severed their connection to "country"; instead, they somehow found ways to live in both the Settler-colonist and the Indigenous Australian worlds. This ability to live in parallel worlds has equipped them with the means to do and think differently. Stephen Muecke points out the potential of an active engagement with such pluralist understandings: "Australians of all complexions agonize over their identities, but let us not leave the only avenue as one of being someone or something. The crucial and nonnegotiable identity emerges in action, or out of inventive resistance, or out of artful politics, it emerges only in movement" (Muecke 2011, 5). Instead of being named as someone or something (Native Title holder, Bachelor of Art, Indigenous custodian), which turns actors into recipients of recognition, Muecke highlights that it is diverse practices that make actants recognizable in the world. It is the music of Archie Roach or Stephen Pigram, as much as the politics of Peter Yu, Patrick Dodson or Marcia Langton, that reintroduce Aboriginality as a central component of what is better termed "transcultural" rather than "multicultural" Australia.

This Australian transculture recognizes the agency of people who feel at home in diverse worlds. Along these lines, it also enables research(ers) to "compare" worlds; that is, "to weigh them, on a more equitable basis" (Latour 2013a, 21). The opportunities offered by related collaborative economic development are far more sustainable than ambiguous calls for industrialization without much definition provided as to where or in what (let alone for whom: fauna and flora, a local population, an industry with headquarters elsewhere, global shareholders?) such 'opportunity' resides.

A collaborative understanding weighs the values of people and land on a more equitable basis, which contests the "opportunity" of industrialization projects and how proponents bring it forth. My ethnographic data show how, despite a long history of Settler-colonialist intervention and its negative environmental and cultural consequences, some Indigenous and non-Indigenous people have avoided subordinating their culture and heritage to it, despite misrepresentation and other constraints, as the anonymous former employee of the Kimberley Training Institute explains:

> *Interviewee:* If you come from a mainstream society and your uncle runs the local shop and your auntie runs the supermarket and your cousin runs a petrol station, you know a little bit about business. And you have a network of contacts to ask. But if everybody who surrounds you is on welfare, the only thing you know is welfare.
>
> *Carsten Wergin:* . . . which is not necessarily a problem only for Indigenous people, or just in Australia . . .
>
> *I:* It's a problem everywhere, absolutely.
>
> *CW:* Once you grow up in this context, then for you that's what the world is like.
>
> *I:* That's right! So what we tried to show people was a broader world. And once we had some operating business systems, then all of a sudden you get that bit of a knock-on effect. But from a government perspective, they want a knock-on effect in 10 weeks. But this is generational. It's taken 70 years, a hundred years, for people to get to this stage. . . . People were here minding their own business and then non-Aboriginal people come in and said, "Okay, you mob, you're all going into missions. You mob, you're going to get this. You mob, this is what you've got." To expect it to change in a day or a week or a month or a year is ridiculous (. . .) And then they go, "Oh well, blackfella's hopeless, he can't get organized." That was the other thing people used to say about blackfellas, "They're unreliable, they can't get organized . . ." I said to people, "You ever been to the football field on a Saturday, you ever seen the Peninsula Bombers [Dampier Peninsula Football Team]?" They all get in a bus and they all come down, and if the game starts at 2:00, they're all there dressed and ready to go at 2:00.
>
> *CW:* That's a good example, because what I hear a lot when I talk to people in tourism is, "One of the problems is, we want to organize a tour but then they're not there."
>
> *I:* Sure, sure . . . Because people don't have the same commitment to a job that we have to a job, because it's a different culture, it's a different idea of what's important (. . .) I get out [to the community] and I go (. . .), "We're going to give this training today, okay?" And they go, "Look, grandma's sick. She needs to go into the hospital" (. . .) So we get grandma, we put her in a car and we take her to the clinic. So while we're waiting at the clinic, the fuckin' tire on the car goes down (. . .) So then we go, "Where's the spare?"

> And they go, "Well, we haven't got a spare. We'll have to get this one fixed." "Okay, where's the jack?" "We don't have a jack either." So we need to get a car to get a jack, to jack it up, to take the tire off, to get the tire to get it fixed.
>
> So (. . .) we got the tire off and we go, "Where can we get the tire fixed?" And they go, "At the garage over there." "Cool, let's go there." "Oh, it's closed for lunch now. It's not open until one." "When's grandma going to get out of the clinic?" "Grandma gets out of the clinic at two." "Okay, well . . . . Then we may as well go have lunch." So we go have lunch and do something. "Why don't we go fishing?" Because someone comes past in a car and they go, "Well, let's go fishing and then we'll come back and then we'll fix the tire and pick up grandma." So all of a sudden, the day is done. And that's one day! (Transcript 121122_KTI)

To reconcile such different world(view)s might be the greatest challenge of all for Australia's contemporary transcultural society. The *Lurujarri Heritage Trail* has been an exemplary means of such reconciliation. It is an economic activity, but one that is done rather than spoken of. It brings people and place together through active participation and the experience of collaboration.

One anonymous participant in the 2015 Trail, a non-Indigenous woman in her mid-forties from Sydney, explained to me her interest in such world-making experiences and the *Lurujarri Heritage Trail*. It is a longer quotation that is worth including here because it also summarizes the argument so far. The interviewee begins with a conventional Settler-colonialist dualism of "us versus them" and the related fear that Aboriginal knowledge is in danger of being lost. Then she turns to the qualities of walking the land, as opposed to other means of travel, and the way in which this has helped her overcome the dualisms she started out with:

> As little kids, we spent some time in the country but that's just because my father came from the country. But I'm too disconnected from Indigenous Australia and I just think they've got so much wisdom that we can learn from but we've actually been blind too. But hopefully, we and they will catch up just in time so that it won't be lost. So that was certainly part of the attraction... And just getting back to nature (. . .) sleeping out, looking at the stars, washing in the ocean (. . .) getting back to as real life as you can (. . .) All those things, physical and personal... Getting back to nature, trying to become a little bit closer... But also support (. . .) the Indigenous Australians. There are so many problems associated with the Indigenous Australian population... Some of them seem so intractable and so difficult that you almost give up (...) And I think sometimes people just think too hard. . . . I think it's fantastic to see any Indigenous community successfully running (. . .) tourism things, but also this added value. . . . That they are teaching their youngsters about their traditions and about the Trail and about the way of life. . . . It's just a much more connected way of life than we've seen civilized [*sic*] people end up living . . . .

> For years, we would look at people who are Indigenous as those who had lived simpler lives, less socially evolved or whatever. But over time, I think a lot of people are realizing that they are closer to the truth or closer to living in harmony with the world than we are (. . .) They are not driven by a set of tasks and timetables and, you know, "We are going to do this and we are going to do that." It's just like, "Yay, it will happen, just don't get too caught up with the details." (Transcript 150901_anonymous)

What the participant highlights at the end of this interview excerpt is a general desire for what she first describes as a wish to reconnect with nature. She admires this in her Indigenous hosts, who appeared "not [to be] driven by a set of tasks and timetables." The Trail participant found inspiration in the very opposite of what Settler-colonialism relies upon. Not rationality (ownership and naming), a teleologically minded futurism, or an understanding of humans as masters over the environment was what she cherished from the Trail experience, but its "otherwise."

To sum up, what I have argued so far is that such an "otherwise" is found in world-making activities like the *Lurujarri Heritage Trail*. These assemble diverse people and their world(view)s in onto-epistemic partnership with an ecological foundation in the land. The land is a resource, but one that is valuable in the here and now, and for very different reasons than those ascribed to it by late-liberal market economics. How to think this "resourceness" of the land, and how it is ontologically different from the resource industry, is the question I turn to now.

4.1 RESOURCES AS ASSEMBLAGE

Since the beginning of Settler-colonialism, the promise of economic development through the exploitation of natural resources has been considered sufficient reason to relocate population groups and to legitimize the destruction of vast landscapes and the pollution of the seas. A poignant example of the unforeseeable consequences this can have arose in mid-November 2015, when Brazil's mineral-rich state of Minas Gerais was hit by an environmental catastrophe: following the collapse of two mining dams, the town of Bento Rodrigues was flooded by 60 million cubic meters of toxic mud. The iron-ore residue then continued down the *Rio Doce* ("Sweet River") and polluted the water supply of hundreds of thousands of residents before finally making its way into the Atlantic Ocean.[4]

Bruce Douglas noted in *The Guardian* on 13 November 2015 that the then president of Brazil, Dilma Rousseff, compared the scale of the damage to the 2010 Deepwater Horizon oil spill in the Gulf of Mexico. She termed it "possibly the biggest environmental disaster to have impacted one of the major regions of our country." By this time, at least fifteen people had lost their lives, entire species had been made extinct, and biologists estimated that it would take a hundred years for the remnants of the catastrophe to disappear.[5]

More often than not, these threats that stem from large-scale resource exploration projects in countries like Brazil, Mexico, or Australia, and their failures, have hazardous consequences particularly for Indigenous populations. This changing perception of the environment as something not only subjected to but also determined by human action has become central to much academic and non-academic discourse. It is generally associated with the three global crises in the areas of economic downturn, growing cultural conflict, and climate change in an already "overheated" world (Eriksen 2016). "Hyperobjects" (Morton 2013) like climate change are a global concern that cannot be tackled by one nation-state alone. They thus challenge spatial practices such as border constructions. This sparked a new debate about no less than the future of humankind in general, for example in regard to the relocation or managed retreat of people living in coastal areas, whether those are large cities or remote tropical islands. This again turns attention to Indigenous knowledge and practice as alternative sources of knowledge and know-how that had been ignored in favor of Modernist attempts to tame nature for the purpose of economic gain.

For me, the Aboriginal concept of "living country" takes the role of a "theory machine" in this regard (Helmreich 2011). It offers an alternative perspective on the transformation processes triggered by the three global crises. The argument I put forward is that these crises are conventionally used to articulate a concern of potential loss. However, more important is their rendering of modest forms of biocultural hope (cf. van Dooren et al. 2016), through the re-emergence of an "otherwise" that offers new collaborations, and alternative lifestyles and forms of engagement with the environment.

The three crises of globalization (economic, cultural and environmental) are intrinsic to the conflict over the industrialization proposal for Walmadany/James Price Point. What motivated the majority of the proposal's opponents was the growing awareness that exploration initiatives entail incalculable risks and unforeseeable consequences for people and place, heritage and culture, fauna and flora, land, and sea. Resource explorations remain central to both governments and industries around the world ensuring prosperity and particularly so in Australia. But the violence of these efforts can no longer be downplayed or ignored. A considerable number of anthropological researchers highlight the advancement of state-sanctioned multinational companies into rural areas and onto Indigenous land (cf. Richardson and Weszkalnys 2014, 5), while oil production in Indigenous communities has been a particular concern

(Behrends 2008; Jorgenson 1990; Mitrofankin 2006; Morrow 1994; Roon 2006; Stammler and Peskov 2008).

Over the past twenty-five years and more, anthropology has been at the forefront in studying resources in light of their "resourceness," that is beyond their mere status as commodities (cf. Richardson and Weszkalnys 2014, 6; also Rogers 2015, 375).[6] For example, Richardson and Weszkalnys (2014, 7) understand resources in relation to broader social science debates about materiality. This for them leads beyond a conventional analysis that sees research on extraction industries linked solely to questions of techno-scientific practice.

Adding to this attempt to broaden the understanding of what a resource is are the contributors to the book *Subterranean Estates: Life Worlds of Oil and Gas* (Appel, Mason, and Watts 2015). Their writings emphasize the need to study the collectives that assemble around the resource sector's interventions in the environment, as well as the social fabric of local communities living close by, particularly in regard to alternative organizations of social and political relationships that are triggered by such interventions by the industry.

Drawing on these and other works, I seek to alter the use of the term resource, moving away from Modernist understandings led by late-liberal market economics and toward alternative values of the land and sea. Clearly, resource exploration initiatives continue to impact all aspects of human and more-than-human life in communities and environments around the world, sometimes in ways that are yet to be accounted for. They modify people's sense of belonging, particularly in places such as Australia where an Indigenous population has historically lived in an entangled relationship with the environment.[7]

The analysis of related modifications to environments and communities that are triggered by resource exploration initiatives is obscured by the four pillars of Settler-colonialism, and in particular by a Modernist timeline that focuses on "opportunities" and progress (but rarely pitfalls) of exploration projects. Such teleological approaches are ill-suited to representing the diverse consequences of economic, political, cultural, and environmental interventions in other people's lifeworlds. They render invisible the alternative visions for entangled futures that opponents of such projects provide when they emphasize the qualities of the present over potential future gains.

The clearing of forests, the planting of fields, the building of roads and cities, the extraction of minerals, the excavation of ocean floors, and the damming of rivers provide some evidence of the possible transformations a local community can face. Meanwhile, the proposed shift toward "resourceness" highlights that communities and people are not passive victims of such expansionism. Existing research proves quite the opposite (Gilberthorpe 2007; Jorgensen 2006), namely the potential for active engagement by diverse stakeholders, including local communities, in related decision-making processes. The question remains, however, whether at the end of the day their voices will be heard.

How voices got to be heard is the question I pursue in my ethnographic analysis of the successful opposition to the industrialization proposal for Walmadany/James Price Point. These voices were raised because of the "resourceness" of the place and its people, whose values were assessed outside and beyond the economic mainstream. As a result, the potential economic benefits of the industrialization proposal were not the central concern of protesters. Instead, they drew attention to the qualities already in place. Simple slogans like "We love Broome," paraded by mothers with babies on their arms during Broome's 2012 "Mother's Day Protest" were one such means by which the community reclaimed its discursive power over the conflict (figure 4.1).

The Intimate qualities of such protest action sit in stark contrast to governments and industries, which continue to legitimize industrialization projects and related ecological transformations based solely on the results of natural science research and abstract economic modeling. They make a good case to highlight that land and sea are not understandable solely in Western terms of their size, utility, value, and risk—not least because

Figure 4.1. Protest Action on Mother's Day 2012 in Broome. Photo: Carsten Wergin.

the contemporary global crises are effects of such normative assessments. This casts serious doubt on what has been an amalgamation of Indigeneity with marginality, since in many places affected by the resources boom, such as Fiji, Papua New Guinea, Timor Leste, and the Canadian territory of Nunavut, Indigenous people constitute absolute majorities and dominate political processes (cf. Weaver 2016, 30).

Again, the mainstream economic model seeks to dominate the discourse by focusing questions of potential benefits solely on the economic value of the resources that are to be explored. Meanwhile, localized groups can find powerful means of opposing such industry attempts, if they are able to reclaim the discourse and draw attention to the "resourceness" of a place: the values that are already in place. To give full recognition to these diverse views with regard to the value of resources is essential (Kirksey, Shapiro, and Brodine 2013; Tsing 2015a; Wergin and Muecke 2012). In this way, resources become more than a means to trace social relations. They become the source of an alternative reading of sociality.

A focus on socialities ties back to the Indigenous perspectives on the environment outlined in the previous section. It emphasizes the multifaceted entanglements of people and place both within and beyond resource exploration initiatives. In these settings, neither communities nor the environment can be termed passive recipients of economic and political decision-making, not least because mainstream economic interests are often not their main concerns. My ethnography highlights the resourceness of Walmadany/James Price Point, Broome and its adjacent region in this way—not from the perspective of market economics that emphasize what might be, but based on the question of how the place *assembles* multiple actors who seek to preserve it, and for what reasons.

Philosopher Gilles Deleuze and psychoanalyst Felix Guattari (1988) introduced the concept of "assemblage" in order to describe a horizontal network of materialities across a number of domains. Instead of the emphasis on a fixed ontology, the authors define social formations as being composed of heterogeneous, relational elements. This assemblage theory is backed by an Indigenous world(view) that emphasizes the entanglement of people and place, land and sea. Similarly, "new materialist" discussions of the interstices created by human and more-than-human actors highlight diverse elements of an assemblage as "instrumental in making the conditions of life visible, factual, and politically operable" (Bessire and Bond 2014, 447; see also De Landa 2006).

Assemblage theory is helpful in avoiding what Bruno Latour calls metadispatchers (such as "modernity") that divert research away from everyday experience and into abstraction (Latour 2013a). I approach the resourceness of Walmadany/James Price Point in a complementary way: as something that is defined by overlaps, networkings, and knots (Ingold

2009). The focus on assemblages thus underscores the logic of systems and other Modernist reasoning (cf. Haraway 2015; also Clifford 2013; Moreton-Robinson 2003). In light of this, my inquiry presents what Arjun Appadurai (1986) famously termed competing "regimes of value." In my case, *Goolarabooloo* regimes of value attribute to "living country" a living heritage, and particular qualities of *Bugarrigarra* that stand against those of the resources industry, which not only promises economic and social benefits in Australia but also for more than two centuries has been the economic pillar of the nation-state.[8]

Meanwhile, mining generates enormous wealth for shareholders, corporations, and governments. However, it puts significantly fewer people in employment than, for example tourism (see above). And, in doing so, it runs the risk of undermining the actual and potential diversity of local economies. Therefore, what deserves considerably more attention than merely a focus on its economic relevance is how the resource industry continues to short-change the desires, beliefs, politics, and aspirations operating at local community levels.

My ethnography of the resourceness of Walmadany/James Price Point pushes into this research gap. I set out to study the *Lurujarri Heritage Trail*, which turned out to be not a conventional staging of Indigenous culture for non-Indigenous participants but the source of a powerful opposition movement in an unprecedented "mining versus tourism" challenge.

4.2 RESOURCE EXPLORATION VERSUS HERITAGE TOURISM

In the course of my fieldwork, the *Lurujarri Heritage Trail* turned out to be significantly more than an Indigenous tourism experience. It was the source of, and motivation for, multifaceted forms of resistance against the industrialization proposal for Walmadany/James Price Point. A second existential conflict along similar lines is therefore the subject of this ethnography: that between the two largest export industries in Australia and the world, mining and tourism. My analysis is informed by both the competing world(view)s of Settler-colonialism versus Indigeneity and the values of the iconic tourist town of Broome, its pristine surroundings, and particular social fabric that were to make way for the installation of a $45 billion AUS Dollar LNG facility 50 kilometers north of it.

The conflict over Walmadany/James Price Point is a prime example of how the diverging values of the mining and tourism industries have turned into a "wicked problem" in many places around the world (Wergin and Muecke 2012). Mining developments of the nineteenth and early twentieth century meant the construction of new towns and homes for miners and their families. Today's industry expansionism, however, is dominated by a fly-in, fly-out (FIFO) workforce and a destructive, rather than constructive, effect on local communities that leaves towns deserted and/or repopulated with what caters to the needs of a predominantly male workforce, from bars and pubs to sex workers (Carrington et al. 2012).

In remote Australia, the mining boom forces leisure tourists to compete with this FIFO workforce for limited accommodation and flights. All sectors of the tourism and service industries are thus affected by resource extraction economies. Aviation, food services, and accommodation providers might benefit from a significant increase in business tourism but struggle to keep up quality and staff numbers as it becomes difficult to compete against the wages offered by the mining industry. Meanwhile, the social value of mining bears little resemblance to the effects of resource extraction and profiteering on a community and its social fabric. For example, despite the need to improve social well-being in remote communities, only an estimated 0.5 percent of mining profits are allocated to Indigenous people, while the mining industry employs about 150–200,000 FIFO workers every week (Carrington et al. 2012).[9]

Tourism in its various domains can be equally exploitative of local environments and communities. However, in comparison to mining, tourism may also promise more environmentally and socially sustainable outcomes, while those parts of the tourism sector that value other than profit-driven gains, such as Indigenous, environmental, and heritage enterprises, experience considerable competition from mining. A tourism product like the *Lurujarri Heritage Trail* is linked to values "embedded" in people, places, stories, landscapes, and heritage sites. As such, it fulfills the main expectation put on Indigenous tourism products and destinations. These advertise the experience of distinct socioeconomic systems, strong ties to particular places, environments, and their heritage, and a continuity of pre-colonial and/or pre-settler traditions that are linked to a commitment to sustain those (cf. Weaver 2016, 29–30; see also Sofield 2002).

Contestations of such values of Aboriginal "country" that are on the one hand central to the "embeddedness" of Indigenous people in the land, and on the other what make the Indigenous tourism industry so attractive for many of its customers, have turned into yet another "wicked problem," this time for social science research(ers) and policy-makers, because the immaterial and material values they transport and protect through the immanent relationship between people and land are confronted with the radically different values, weight, and force of commodities, technical and financial infrastructures of mining.

The conflict over Walmadany/James Price Point is no exception, since here as well a lot of time and effort is spent to determine the rightful stakeholders. Some of them are human, such as Aboriginal Traditional Owners (TOs), local community members, small business owners, or a general public that is meant to profit from the mining boom. Other stakeholders are non-humans, such as bilbies and whales, or rare turtles and dinosaur footprints. The "wickedness" of the problem lies in the question: How does one come up with a decision-making process that brings all of these together?

For example, the iconic image of Broome's Cable Beach, with its clean yellow sands and spectacular sunsets, is of undeniable value to locals and tourists but found itself in competition with the fires of an industrialization proposal that would illuminate the night sky and that brought with it ship movements and general pollution that would be visible all along the beach. The attractiveness of the beach, for tourists and travelers as well as the local population, which the journalist Astrid B. outlined vividly in her earlier quote, would plummet in value, unless people prefer to see the sun set behind an LNG vessel rather than the horizon.

Meanwhile, the industry claims to generate wealth and new jobs in its own terms: new medical facilities for remote communities, better access to education, and promises of jobs in the mining sector for the younger generations. What is unquestionable in this regard is that the wealth of mining magnates continues to increase substantially, but in contrast, the well-being and general quality of life in local communities and their environments diminishes.[10] Indigenous people figure particularly high in unemployment, alcohol and drug abuse, domestic violence, and crime statistics, all of which have been a feature of life in remote parts of Australia. In contrast to this, the chapter set out with the example of how a growing Indigenous tourism sector has the potential to unleash a revitalization of Indigenous world(view)s in a more organic sense, based on new collaborations that stem from tourism experiences like the *Lurujarri Heritage Trail*.

Indigenous tourism is a rather recent discovery compared to the introduction of mainstream tourism to Broome. The latter really took off with the opening of the first five-star resort, Cable Beach Club, that was built in the 1980s by the late Lord Alistair McAlpine, former treasurer of the British Conservative Party and advisor to Margaret Thatcher. With the decline of the pearling industry, McAlpine became the key figure responsible for the development of the town's tourism industry. He rescued Broome from bankruptcy with his investments and stimulated tourism through the construction of Cable Beach Club, a crocodile park and zoo, and another fifty-odd property purchases, notably historic homes of master pearlers that today have become part of WA heritage listings.

Elisabeth F., whose family is closely linked to one of the largest and oldest pearling companies operating in Australia, describes his impact as follows:

> *Elisabeth F.:* I started coming down here (. . .) in the 1970s (. . .)
>
> *Carsten Wergin:* It must have looked a lot different.
>
> *EF:* Very different, yeah. Cable Beach Club was a nudist. . . . You know, was just a whole lot of backpackers.
>
> *CW:* Nudist beach.
>
> *EF:* Yeah, there was nobody here, you know? And it was just hippies living down there. And all the master pearlers' homes were (. . .) becoming derelict until Lord McAlpine came in the 1980s and fell in love with the town, fell in love with the *pindan*, fell in love with Cable Beach, suggested to the local government and the state government that they give him the land on the beach, and in return, he'd build a caravan park (. . .) Because basically it was an illegal caravan park on the other side. So, he suggested that it be formalized. So he bought that caravan park; and he built Cable Beach Club, which *transformed* the profile of the town (. . .) and the architecture and the future vision for the town. And based on preservation, actually, rather than not caring about anything. . . . McAlpine bought most of the old master pearlers' homes and restored them all, sold them to friends and kept some himself. If he hadn't, the whole town would've become a besser-block town. Everything would've just become red brick, closed off, air-conditioned, mundane, ordinary . . . ugly . . . with a beautiful beach . . . Ugly with a beach. (Transcript 120504_MaPa)

In 2015, a bust was erected in honour of Lord Alistair McAlpine not far from Cable Beach Club (figure 4.2). At the time of his intervention, the one road that connected Broome with the rest of WA was still a "dirt road" for about 600 kilometers, until Port Hedland. During the "wet season," between November and April, it remained subject to extensive flooding that can still hinder food trucks reaching the town's supermarkets, even though the road is now bituminized. Even in the 'dry season,' between April and November, the old road was difficult to master with cars other than four-wheel drives.[11] Of course, transport with legendary "Holden utes" and other unsuitable vehicles was frequent, but so were the accidents with wildlife (kangaroos, camels, donkeys, and horses), flat tyres, and overheated engines.

From the day the road was sealed in 1981, things began to change for Broome. The town is still flooded by rain in the wet, but more importantly by camper vans in the dry—so much so that so-called overflow caravan parks are introduced in places such as the McAlpine oval (figure 4.3). This larger space near the town center was intended to offer children sporting activities within the town precinct. Despite the significant economic

Figure 4.2. McAlpine Statue at Cable Beach, Broome. Photo: Carsten Wergin.

benefits that tourism has brought to the town, its social impact remains visible and a reason for Broome residents to criticize the growing influx. Some have even termed tourists "terrorists," responsible for long waiting hours at the post office, annoying questions about Broome's history, and the crowding out of restaurants and shops. However, the "tourist dollar" was appreciated and the local economy gained strength after the steady decline of the pearling industry.

George H. is the owner of Astro Tours, a stargazing venture. He is also famously known around Australia as Space Gandalf, due to his 2017 appearance on the ABC television programme *Stargazing Live* with his Kimberley hat and long white beard. Below he gives another account of the transformations that Broome went through at the time when McAlpine began his investments, and when the main road was bituminized. He describes with insightful detail how he arrived in Broome, settled in, and began to make a living as one of the last pearl divers:

> *George H.:* I hitchhiked, February 1982 (. . .) I've always been looking for places with freedom of thought. And that's what I found in Broome (. . .) Didn't matter whether you were black, white, or brindle, people would accept you and you could just be who you are.
>
> *Carsten Wergin:* Did you hear from somewhere that Broome would be like that or was it a coincidence?
>
> *GH:* I guess a little bit of both. As a Western Australian, growing up in the southwest of WA, you would see Broome in the schoolbooks as this place with pearl luggers. . . . But it always seemed like this far off place that you would never dream of going to.
>
> *CW:* Even for a Western Australian?
>
> *GH:* Yeah (. . .) Western Australia's so big. It's nearly 3,000 km from where I was born (. . .) Yeah, and one day I found myself just heading in this direction without really knowing why. . . . And I arrived in Broome and didn't really (. . .) gel with the place straightaway. I wandered around and looked at things and wondered if I was going to stay. . . . And then one day I'm walking down the street in Chinatown and I walked past a pearl shop and something in my brain said, "I wonder if they want any divers?" Because I'd been a scuba diver since I was fourteen. . . . So I walk in and said, "Do you guys need any divers?" And they said, "Yes, start tomorrow" (. . .) Well, on the same day I was walking around behind the sand dunes here [in Miniyrr Park] and I found a shack that was empty and I moved in (. . .) So, yeah, exactly a week after I got here everything fell into place for me. (PAUSE)
>
> So, I was diving in '82. It was maybe six years out of [hard hat diving] (. . .) But you know, even though we were using skin diving gear it was still pretty rough and ready days. And we didn't use dive tables and I was a pearl farm diver, so we'd do maybe six dives a day (. . .) It was a lot of repetitive up and down diving. . . . What stopped me eventually was I did get bent a couple of times (. . .)
>
> *CW:* Did you already have tourists? People that would visit the farm?
>
> *GH:* Well, no. . . . But I mean Willie Creek has been set up in the beginning as a tourist thing. It isn't a pearl farm at all. They bring pearls in from a farm, they put them in the creek and they call it a pearl farm. . . . It was only ever a tourist thing. (Transcript 130314_GQ)

George H., in this interview excerpt, describes the daily routine of a pearl diver in the early 1980s, at a time when mainstream tourism was only beginning to populate Broome. Pearling, pearlers, and pearl luggers were both real-life activities and legendary stories told in schoolbooks, but not yet the "tourist realism" they were about to become (Bruner and Kirshenblatt-Gimblett 2005). Once the road was sealed and travel became more comfortable, their exotic appeal stimulated more people to visit the

Figure 4.3. McAlpine Oval in Broome Converted to Overflow Caravan Park. Photo: Carsten Wergin.

town. George H.'s account thus ends rather abruptly with what became of pearling once the "tourist dollar" was discovered as the new resource to be exploited in town. It led to the construction of Willie Creek Pearl Farm, which has become one of the main tourist attractions in Broome, but which, as George H. notes, "isn't a pearl farm at all."

In Broome, the pearling legacy is manifested in such tourist realism: a place that calls itself a pearl farm, but only came into being because of tourism. In addition, local museums retell the history of pearling, and Pearl Luggers Café and Bar, another venture owned by Willie Creek Pearls, caters for evening functions and weddings with two original luggers on scenic display (figure 4.4).

This transformation shows the unlikely resemblance of tourism to the resource industry. Both have a lasting impact on the social fabric of a town, while also depending on larger infrastructures such as sealed roads and port facilities in order to function properly. Aside from the sealed roads that made tourism in Northwest Australia increasingly popular, air travel is another significant component in this regard. In the 1980s, Qantas operated its first regular domestic services to Broome via Darwin and Perth. Meanwhile, another competitor, MacRobertson Miller Airline (MMA), which was known among locals only as "Mickey Mouse

Figure 4.4. Broome Reggae Band The Mexicans Setting Up at Pearl Luggers Café. Photo: Lina Pranaitytė-Wergin.

Airlines," went into bankruptcy after a pilot strike in the early 1980s. This also impacted McAlpine's investments, and he ultimately had to abandon his projects in Broome because of significant losses.

At the time, air linkages such as those of Trans Nusantara Airways already tied tourism and the resources industry together. Its planes were used by Burmah Oil Co. to move workers between Timor, Bali, and Broome. One of those ran off the runway on 21 January 1974 and was later used as the first Broome Tourist Bureau.

Today, the Broome Visitor Centre is located adjacent to Male Oval, at the entrance to the town. All regional buses stop there and most travelers who recognize Broome as "the gateway to the Kimberley" seek advice from it, whether they arrive in large recreational vehicles (RVs), as backpackers, by bus or plane. Broome's mainstream tourism industry is well established, both geographically and mentally. Even the global financial crisis (GFC) of 2008 has not left a significant mark on the town. On the contrary, the Bali bombings, tsunamis, and the GFC drove more Australians back to domestic travel. Furthermore, Broome's diverse "exoticizing" qualities are attractive and set it apart from other Australian destinations. The RVs and campervans will continue to block roads, while their owners will continue to cause longer waits in the post office and be made responsible for the frequent sell-outs of finger foods, bread, and nibbles in the town's two largest supermarkets, Woolworths (Woolies) and Coles.

One participant in the *Lurujarri Heritage Trail* of 2015, a non-Indigenous woman in her mid-50s from Sydney, visited Broome briefly for a couple of days before and after the walk. When we met a few months later, she identified two further prominent qualities that set Broome apart, beyond the pearling legacy and Cable Beach. The first, the type of people, underlines the initial points made by George H. on what attracted him. The second concerns the architecture of the town and links up with the accounts given by Elisabeth F.:

> [Broome] seems to be a place that welcomes people who are a bit different, eccentric even, which I think remote places often are (. . .) And that's a nice quality, I think. The other thing I like is that the building material they've chosen is the typical one . . . corrugated iron. The other thing is, if you walk down the suburban streets in Cable Beach, the houses look like they are on steroids. The blocks are really big and the houses are like three times as big as a big house here [in Sydney] (. . .) That's probably Broome. They didn't seem to have any regulations. . . . In Sydney, you have to have a guttering or you have to have a metal connector from the roof to take the water. . . . But in Broome, they didn't even have guttering because (. . .) in a torrential rain, there is no point (. . .) So they just make sure that the roof goes far enough away from the wall. Then it's not a trouble (. . .) So I like that architecture that was quite specific to Broome. I think it's an interesting town, interesting place. (Transcript 150901_anonymous)

Apart from the interesting mix of people and distinct architecture, another significant factor that makes Broome a sought-after destination for older domestic travelers is medical services. This was brought up by Margaret and Ron, an elderly couple from Beecroft near Sydney whom my family and I befriended on our very first flight to Australia, and whom we encouraged to housesit our apartment in Broome while we were away:

> *Ron:* As you know, I'm a bit older, and I always look at where I'm going as to my access to medical facilities. And I had an incident in Broome and I needed medical attention. And I was horrified that . . . first I thought, "Oh, I'm in Broome, where's this going to take me?" But we went to a doctor that you recommended and she was good, absolutely good. But she was leaving to go on her long weekend holiday. She got me to go to Broome Hospital (. . .) and we were blown over by the treatment we got there. It was *excellent* and the attention they paid to us (. . .)
>
> *Margaret:* We also found out, because we were such regular visitors to the hospital, that people are quite well paid and rewarded . . . for taking a job (. . .) They get an apartment provided. . . . A huge subsidy of their electricity, a fully furnished . . . cutlery, crockery, the whole thing (. . .) And we spoke to a number of youngish people who had come there in particular to nurse,

just to save money (. . .) The first thing I would ask people when you got into a conversation was, "Do you live here?" "Are you from Broome?" I could count on one hand the number of people who were (. . .)

R: We met a few . . . nurses like that as well who said, "I might spend two or three years here to get some experience . . ." Said, "It's interesting also to work partly with Indigenous people. It's a different kind of approach in . . . the work and everything . . . makes it more interesting . . . ." But then, "I'm planning to leave again" (. . .) Especially when they are at the point where they want to have children and family and stuff . . . . (Transcript 130723_1_RandM)

Despite the qualities and particular details that set Broome apart and make it attractive for a wide group of people, from work and travel visa holders to retirees, the continued success of the town as a tourist destination and the gateway to the Kimberley was being questioned at the time of my fieldwork by the then Shire president Howard R., when he told me in a personal interview in 2013, "tourism won't save Broome."

What put the town in jeopardy in his mind was a significant problem that was again related to Australia's resource extraction economy. In 2011, Australia suffered from what in economics is called the Dutch Disease.[12] The value of the Australian dollar was so high that it became significantly cheaper to take a holiday overseas than to do so within Australia. According to Howard R., the economic antidote to the decline in visitor numbers was therefore not to be found in further diversification of the tourism industry, but in the introduction of a whole new industry. The long-term vision had become that mining would replace tourism, just as tourism had replaced pearling some forty years earlier. Resource exploration was to make a rather enormous entry, in the form of the $45 billion AUS Dollar LNG facility 50 kilometers north, at Walmadany/James Price Point.

Histories of pearling, tourism and the LNG proposal are therefore tied up in my ethnographic inquiry into the struggle of the *Goolarabooloo* and their collaborators to preserve their "living country." The final section of this chapter will return to the *Lurujarri Heritage Trail* to illustrate in a more graphic sense "the ground" on which this inquiry rests. This introduces the values of a "living country," whose recognition has been a game-changer for both the successful protest movement and the new Indigenous tourism enterprises that emphasize Aboriginal culture and heritage already in place.

4.3 AUSTRALIAN INDIGENOUS COUNTRY AS LIVING HERITAGE

In the previous sections I discussed how over the last one hundred years, Settler-colonialism has profited from the various qualities of Broome and its adjacent region. Those have included the rich supply of *Pinctada Maxima* mother-of-pearl shells and pearls. The pearling legacy continues to exist, but its visibility has been reduced to shops, museum spaces, and tourist realisms like Willie Creek Pearl Farm, which have been designed as heritage products to extract the "tourist dollar" from a growing number of visitors.

At the time of writing, the resources industry remains poised, as it was with the LNG proposal for Walmadany/James Price Point, to override the qualities of the Kimberley, in particular those linked to Indigenous Lore and culture, by advocating possible economic gain, this time through use of the controversial technology of fracking and sand mining. These particular qualities that are emblematic of the lived experience that the *Lurujarri Heritage Trail* offers its participants each year have been foundational in allowing the opposition movement against the LNG facility to succeed because of their emphasis on the entanglement of people and place.

Such interconnectedness of people and place as the foundation of Aboriginal Lore and culture has of course been outlined before, for example by Deborah Bird Rose in her influential essay *Nourishing Terrains* (1996). Stephen Muecke has made further significant contributions to this in his books *Ancient & Modern* (2004)—in which he considers possibilities for the development of an autochthonous Indigenous philosophy as part of modern Australian life—and *Reading the Country* (Benterrak, Muecke, and Roe 1984). The latter is particularly relevant to my project, because it was coauthored by Paddy Roe, the late *Goolarabooloo* Senior Law Boss who initiated the *Lurujarri Heritage Trail*.

Such a reading of country demands that people become "nomads" (ibid.): that they experience the environment through movement rather than stasis—as process, personal engagement and interaction, or what Paddy Roe and the *Goolarabooloo* community call "living country:"

> Where the land is whole and complete; where the interaction between people and land is alive through law and culture; where the spirit of the land is "standing up" and vibrant. Living country differs to [*sic*] cultivated country, which is where there is cultural confusion; the land has been replanned one layer over another; the spirit is hiding, withdrawn, waiting. (Roe and Hoogland 1999, 30)

Participants in the Trail are taught about "living country" and its living heritage through stories that invoke the relationship of its custodians to the land. They also learn how to interact with it themselves—when to go fishing, how to collect bush honey, and make boomerangs and clapping sticks. They are shown culturally significant sites and, after a few days of walking and connecting with the environment, experience what it might mean to be able to "read" it. To highlight this, in Australia, both

Indigenous and non-Indigenous locals speak about being 'on country.' They might live in urbanized environments, but they experience a particular connection to the land every time they go 'out bush' for fishing or camping activities, often together with their extended family.

Today's Indigenous and non-Indigenous members of the *Goolarabooloo* community, notably Frans Hoogland, coauthor of the above quote together with Paddy Roe, continue the *Lurujarri Heritage Trail* with this in mind: the Trail will help to keep country alive, or as Elisabeth Povinelli describes the particular way in which Indigenous people maintain the relationship to country, "the everydayness of their labor-action is swept within the suprahuman realm of a sentient landscape populated with ancestors and totemic beings" (Povinelli 1993, 133).

Describing the landscape as "sentient" aligns Povinelli with the new animists, who according to Nicolas Peterson follow the question of "how to behave appropriately towards persons who are not human: that is, it is about respecting the living world" (Peterson 2011, 169). In the case of the *Goolarabooloo,* such respect is synonymous with responsibility for the living heritage this world entails. Consequently, the motivations for their "labor," for doing what it takes to keep country alive, lie less in expecting a response or responding to more-than-human elements of a 'living country,' than in experiencing responsiveness through mutual responsibility (cf. Bird-David 1999, 77).

In this way, the regime of value that the *Goolarabooloo* adhere to, that is "living country" and living heritage, finds its expression and acknowledgment in experience, which, as Munn has termed it, allows for a "consubstantial relationship between the ancestor and his objectifications" (Munn 1970, 142–43). Thus, being 'on country' automatically means linking up human and more-than-human actors with ancestral beings in what "create[s] the sense of the landscape being almost alive" (Peterson 2011, 176).

Peterson adds "almost" here because in his view the new animists forget that "the ancestral beings and the stories about them are spoken of in resolutely human terms" (ibid.). Meanwhile, as my ethnographic examples below demonstrate, there are aspects of "living country" and living heritage that cannot be spoken about. This is why the *Goolarabooloo* represent the Song Cycle that they are *Maja* (Law Bosses) for not through language alone but in conjunction with experience. Anything else would discredit the entanglement of Aboriginal cosmologies with the world and would downplay the onto-epistemic partnership of people with "living country" and living heritage.

As previously discussed, if taken seriously, this opens up further means of overcoming the false dualism between Indigenous and non-Indigenous people, since country has an equal "say"—for the want of a better term—in determining who qualifies to care for it. Notably, this allowed for Paddy Roe to become Frans Hoogland's mentor in the preservation of "living country" through experience (figure 4.5). Hoogland has become an integral part of the *Goolarabooloo* family. Roe's great-grandchildren call him "Uncle Frans," and they remember him carrying them on his shoulders, sharing stories that are essential to their Indigenous upbringing "on country," while walking the land.

In this sense, the new generation of Indigenous custodians of the Trail has been raised in a collaborative spirit that does not separate Indigenous and non-Indigenous when it comes to protect "living country" and its living heritage. Learning about "living country" helps the *Goolarabooloo* and Trail participants to evade the essentialist discourse of Settler-colonialism, because their engagements with "living country" and living heritage undermine its pillars, and with them its "metadispatchers" such as modernity, progress, and development (Latour 2013a). The latter are commonly used to advocate for industrialization projects to turn "living country" into "cultivated country," to paraphrase Roe and Hoogland. And as they continue, "the spirit is hiding, withdrawn, waiting."

The anthropologist Fred Myers, in his famous studies of Aboriginal Australian acrylic painting (Myers 1989, 1991, 1994, 1998, 2001), illustrates the possible transformation of such Aboriginal inalienable possessions like "living country,"—in his case sacred designs—into commodities, and how this process is fundamentally linked to outsiders, in his case curators, elevating the value of Aboriginal designs in Western terms (cf.

Figure 4.5. Frans Hoogland and Binky. Photo: Carsten Wergin.

Myers 2001, 11). Myers concludes that "the relocation of material culture demands understanding that value is never simply defined but is always involved in global as well as local circuits of exchange, display, and storage" (ibid., 12).

These observations also have relevance for the means in which different regimes of value seek to prevail over "living country" and its living heritage. As Myers points out for objects, the latter two are equally important as "media of social activity precisely because they have properties—because they are vulnerable, fragile, and losable, because they rot or endure" (ibid., 14). Meanwhile, "living country" and living heritage reach beyond conventional heritage discourse about the intrinsic values of museum objects because of the fact that their experience cannot be stored or archived. Both simply present themselves to whoever seeks their experience. Those dwelling in them need to acknowledge that they equally "rot or endure."

No theory about heritage can be more true to Aboriginal Lore and customs than the ones that recognize them as social processes in which the capacity to share or withhold is a marker of social strength and identity (cf. Myers 2001, 15). In this sense, while "living country" and living heritage stand outside of conventional heritage discourse, they ascribe themselves to the study of objectification, of giving form, of becoming (cf. Miller 1987, 81).

Munn (1970) has shown how subjects come into being through their relationships with the land, which Myers interprets as such a type of objectification:

> Country in these situations is an entity that brings with it the capacity for subjective agency, for power, understood as the ability to participate in exchange and in the nurturance of others. Named places, then, are an exemplary case of the inalienable possession, the "dense" cultural object, its meaning generated through its engagement with a complex set of ongoing practices that constitute persons as similar and different in various ways and in different contexts. (Myers 2001, 24)

Relevant here is Myers's emphasis on "ongoing practices," which contradicts the idea of an Aboriginal heritage that could be a static representation of some distant past. Meanwhile, Native Title law or conventional tourism advertisements demand just that: to cultivate a form of "primitivism" that ignores ongoing transcultural exchange (cf. Myers 2001, 42). In other words, rather than override "living country" with "cultivated country" by inscribing it into heritage discourses of modernity and authenticity, the experience of living heritage offers means to access a more nuanced understanding of "country" that is yet to be accounted for in the Settler-colonialist mainstream.

The *Lurujarri Heritage Trail* offers such an experience. It allows for heritage practices to "circulate across boundaries, [focussing] in culture making rather than on relatively static cultural categories" (Myers 2001, 55). The collaborations that formed between Indigenous and non-Indigenous protesters in opposition to the industrialization proposal stem from such experience. They highlight alternate configurations of the political subjectivities involved in the Walmadany/James Price Point conflict, because their shared connection to the land as "living country" fueled a conscious decision to let ecology grow rather than allow economics to "cultivate" the land.

In this way, living heritage plays a fundamental role in the formation of culture and identity, not because it represents the past but because it is an active component in its contemporary experience and future preservation. It provides for multisited engagement, fostering the entanglements of heritage studies, Western science and Aboriginal cosmology. Marala/The Emu Man is a particular case in point. This *Bugarrigarra* story tells about the coming into being of the Song Cycle that the *Lurujarri Heritage Trail* follows as it passes through *Bardi Jawi, Nyul Nyul, Jabirr Jabirr, Ngumbarl, Djugun, Yawuru,* and *Karajarri* country (Tindale 1974; McGregor 1988; Bradshaw and Fry 1989; Stokes and McGregor 2003):

> Three Song Cycles emanate from *Minyirr Djugun Buru* (the greater Broome area): *Dabber Dabber Goon,* which travels east, cutting through Uluru (Ayers Rock, Northern Territory) until it reaches the Pacific Ocean; *Billingun,* which follows the same path as *Dabber Dabber Goon* until it reaches Uluru where it splits three ways (. . .) and *Nunnungurugoon,* which travels along its own path northeast, through the Kimberley. (Salisbury et al. 2016, 2)

In the process of moving through the land and thus creating the Song Cycle, Marala/The Emu Man left three-toed tracks behind. And three-toed dinosaur footprints left by the *Megalosauropus broomensis* are found as testimony to Marala's journey (cf. ibid., 2). They therefore form an important part of the living heritage of the region, while the disappearance of some tracks through natural processes is seen as part of an ongoing unfolding of *Bugarrigarra,* of "living country" and living heritage (cf. ibid., 3–4).

In 2016, Salisbury et al. published a study that identifies these dinosaurian tracks in the Broome sandstone for the intertidal zone in the northern half of the *Lurujarri Heritage Trail* and inclusive of Walmadany/James Price Point, covering approximately 25 kilometers of coastline. Without going into exact numbers, the study found that:

> The diversity of non-avian dinosaurian track types seen in the Broome Sandstone in [this] section of the Dampier Peninsula, each of the main track-

> bearing areas (Kardilakan–Jajal Buru, Walmadany, and Yanijarri) within this stretch of coastline, and many of the area's individual tracksites is unparalleled in Australia, and even globally. (Salisbury et al. 2016, 116)

Beyond the fact that these are remarkable findings, what is relevant for my argument here is that they came about because of the entanglement of living heritage, Western science and Indigenous cosmology, since research(ers) engaged in all three. What Salisbury et al. present in their study leads beyond mainstream heritage discourse, because (1) it merges the values of the past, present, and future, (2) it terms preservation equally existential for ancestral beings, local people and the environment, and (3) its findings, presented in this way, leave ample space for the recognition of alternative, socio-ecological lifestyles beyond a late-liberal mainstream.

In this sense, the protest supporting the preservation of Walmadany/James Price Point against the ongoing expansionism of the resource industry, which after all is regarded as the economic backbone of Australia, can be described as successful because of the living heritage it stood up for and that stood up with it, while the emphasis on "living country" as a common denominator granted the *Goolarabooloo* the final say before any protest actions were taken. The *Goolarabooloo* were the authority and referenced as such by all those involved in the opposition movement. Proponents of the project did the opposite; they made every effort to undermine their authority, up to the point when, on 01 May 2018, Federal Court justice Anthony North delivered the *Bindunbur* and *Jabirr/Ngumbarl* Native Title claims determination for the Middle Dampier Peninsula to the *Jabirr Jabirr*, *Nyul Nyul*, and *Ngumbarl* people, and against the *Goolarabooloo*. Five years after Woodside withdrew from building the LNG facility, "living country" was taken away from the *Goolarabooloo*, as they were not termed its legal TOs (Wergin 2021).

Of central importance was the fact that the court was bound by Western law and therefore focused its efforts on reinstating a situation that had existed prior to 1829, when WA was officially colonized. The *Goolarabooloo* were not the custodians at the time because their community did not exist then in the form that it exists today. But neither did the living heritage they are entangled with and feel responsible for exist then in the form it exists today. It has always progressed in both positive and negative ways, not least since 1829 with the devastating impact of Settler-colonialism. However, recognition of this processual quality of the *Goolarabooloo*'s living heritage was not possible for the court, whose legislation aims to reinstate cultural expression of the past, as if "frozen in time." Below I discuss in detail how the court came to its decision, based on ethnographic data from the final days of court hearings during April 2017 in Perth.[13]

To return to the protest actions that preceded this decision, another strategic choice made by the opposition movement was to link "living country" and its living heritage to the global "mother earth" discourse. This was pertinent during an event staged at Walmadany/James Price Point on 12 October 2011, where Indigenous and non-Indigenous people came together to perform a *Corroboree* (a traditional dance ceremony). One of its organizers, Anne Poelina, provides audio commentary to a video that was produced about the event and that was staged, as she phrased it, "to prevent the crime against mother earth, ecocide."[14]

Such rhetoric resonates well with debates about the predicament of the Anthropocene and similar globalized scenarios about a looming environmental apocalypse. It is interesting to note, however, that despite the awareness of the contribution that Indigenous knowledge about the environment can make, a decision was taken by the *Goolarabooloo* and the citizen scientists they collaborated with to translate their findings about "living country," such as evidence of endangered species, fauna and flora, into prevailing Western science terminology. Proponents deemed it necessary to defeat the work of consultants literally *in its own terms*, in order to guarantee the best result in a court of law. In such instances, both the outcome of the Native Title case and the decision to translate Indigenous knowledge into Western scientific terminology meant a loss of "living country," because it omitted the necessity to experience its living heritage in order to make a qualified assessment of it. Instead, representationism turned living heritage into what might be termed, to again borrow from Roe and Hoogland, "cultivated heritage" that is assessed and documented rather than experienced.

The discussion of objectifications begun by Myers and Miller above ties in with this observation, as does Peterson's critique of new animists who override the values of Aboriginal Lore and customs with anthropomorphisms, in a false attempt to represent what cannot be represented through language alone. Meanwhile, motivations for such efforts are also found in Erich Kolig's (2000, 26) concluding remarks in his article *Social Causality, Human Agency and Mythology*, in which he argues that "Aborigines will soon realise that it is more important and more effective to inscribe their history in books and, thus, on the nation as a whole."

In order to preserve Walmadany/James Price Point, Indigenous and non-Indigenous collaborators chose "books" to present their research results. However, in contrast to Kolig's predictions, I contend that this did not result in "living country" and living heritage losing their significance. A deliberate choice was made not to highlight the importance that both had in achieving more accurate results; well aware of Modernist constraints, none of the collaborators wanted to run the risk of their findings being deemed unscientific. Yes, this means that living country was not properly accounted for. But the shared acknowledgment of "living country" through the active engagement with its living heritage remains the root of collaborations that led to more accurate findings about its significance than those that could be produced with Western science alone.

I therefore encourage readers to hold on to "living country" and whatever living heritage might invoke in them when 'metadispatchers' like scientific evidence, truth or reason seek to override, or rather 'cultivate,' the urge for its lived experience. Willie Creek Pearl Farm makes a good case for heritage taking this ambivalent place. On the one hand, it is a cultivated heritage site produced as 'tourist realism,' and on the other, each year it provides the *Goolarabooloo* with a boat transfer across the creek for Trail participants, making a substantial contribution to the experience of "living country" and living heritage.

McAlpine's Cable Beach Club Resort and Spa is a further example of the touristic "cultivation" of living country. The beach itself plays various roles in this, but also in the lives of Indigenous and non-Indigenous residents. The Native Title holders over the town of Broome, the *Yawuru* people, seek to preserve and/or reclaim parts of it as significant cultural sites. Meanwhile, viewed through a collaborative lens, Cable Beach is also a cultural site for non-Indigenous residents. It holds significant places in the Song Cycle and is significant for Broome locals who cherish it as a place for an early morning walk or an evening dinner and drinks with friends and family, including the free-of-charge spectacular sunset. The beach has further significance also for many visitors to the region. Quite a few of these get married in Broome because of the memorable setting the beach offers, with practically zero risk of rain between April and October (figure 4.6).

Participants in the *Lurujarri Heritage Trail* might pass such cultural performances as they pass Cable Beach Club Resort and Spa on their first day, when leaving Broome north along the beach (figure 2.4). They have their first overnight stop at Millibinyarri/Coconut Wells and continue early on the following day to reach Willie Creek at around lunchtime, cross it and take a break near Willie Creek Pearl Farm. Jeanné Brown, who has been with the *Goolarabooloo* community for nearly thirty years and accompanies the Trail almost every year, expressed to me once how after passing the "farm" the group begins to grow together as a family. People become a community by sharing the experience of leaving "cultivated country" behind.

This image appears romanticizing, but is no less so than what the resource industry and the Australian government have proposed for centuries, which is Australia as a "land of opportunity." As argued above, this opportunity dreaming is nurtured by idea(l)s of development and progress based on the exploration of people and land that are justified by

Figure 4.6. Sunset Yoga and Wedding at Cable Beach, Broome. Photo: Carsten Wergin.

the four pillars of Settler-colonialism. The *Goolarabooloo* and their collaborators present a significantly different view with regard to the values of Australia, namely as a "living country" and living heritage. They emphasize the significance of a place in relation to what is readily available—a reality that is performed in experiences such as the *Lurujarri Heritage Trail*.

"Living country" describes an organic interrelationship of people and place that Trail participants gradually become accustomed to as they learn and experience *Bugarrigarra*, out of which everything emerges and into which everything returns. First, Trail participants meet at the Old Hostel on Dora Street to walk into the dark to Millibinyarri/Coconut Wells, where delicious food and a warm fire have been prepared by the *Goolarabooloo*. In the early dawn, they are offered tea and porridge. It is still cold, so each participant wears numerous layers of clothing. They begin their walk into the day, until the sun and the exercise heats their bodies to such an extent that they peel off almost all of these layers.

When they reach Willie Creek around lunchtime, there are no more scarves, no more jackets, no more long-legged trousers to separate people from country. While they have stripped off Modern amenities on the outside, their insides take longer to adjust. As Jeanné Brown emphasized, internalizing living country gradually begins when people engage with its living heritage and leave "cultivated country" behind. The *Goolarabooloo* have seen this process each time they organize their Trail. They have seen the transformations of people subjected to country for the first time, without romanticizing: simply witnessing people as they stop talking and start listening.

Brown is what might be termed the artist in residence of the *Goolarabo-oloo*. In her drawings, she has documented many aspects of their living heritage—stories as much as particular plants and animals, landscapes, and spiritual places. Her relationship with the community began when she was part of a group of newcomers to the country herself. It was in the early 1990s when, as she described it in a personal conversation, "country grabbed her and never let her go again."

That day we met, Brown tells me about the formation of the annual Trail group in quiet, calm, assured terms. She knows that it will come eventually. It happens every time.

NOTES

1. Personal interview with a forty-year-old male Indigenous collaborator (Transcript_anonymous).

2. An adequate way to address this problem might be to grant significant places, land and sea, the same legal rights as human beings, as has been done with the *Whanganui River* in New Zealand.

3. Personal interview with a thirty-five-year-old female Indigenous collaborator (Transcript_anonymous).

4. In the aftermath of this environmental disaster Brazil's national water agency (ANA) announced that the presence of arsenic, zinc, copper, and mercury in the *Rio Doce* had made its waters unsuitable for human consumption. Furthermore, the lack of oxygen and high temperatures caused by the pollutants had killed off much of the aquatic life along a 660-kilometer stretch of the river, while thousands of hectares of environmentally protected land were similarly destroyed.

5. In addition to the environmental impact, *The Guardian* further reported that the incident had significant political and economic effects. The mine and the dams were operated in a joint venture between the Anglo-Australian mining group BHP Billiton, the largest mining company in the world, and the Brazilian iron ore producer Vale. In the aftermath of the catastrophe, BHP Billiton's company shares lost £8 billion off their value.

6. Jane Guyer opened this field to wider socio-political concerns. She analyzed oil prices across time and space, from negotiations at a single gas station (Guyer 2004) to the national contexts in Nigeria and international negotiations over the Chad-Cameroon pipeline (Guyer and Denzer 2009, 2013). Her work addressed the techniques that companies use to justify high oil prices (Guyer 2009), as well as the spatial, social, and political circumstances under which "the fiction of a single global price for oil is both assembled and challenged" (Rogers 2015, 370). Michael Watts' (2012) analysis of such "oil assembly" further adds to this approach. He moves away from the focus on the close relationship between oil and

the mainstream economy to the study of new forms of chemical technology and knowledge—again, for example, with regard to Deepwater Horizon, the drilling that took place and the environmental disasters it caused.

7. For comparative perspectives, see Carse 2012; Davidov 2013; Rumsey and Weiner 2004; Sawyer and Gomez 2012; Stammler 2005, among others.

8. The "triple crises" in global finance, climate change and oil consumption continue to affect economies worldwide (Papatheodorou, Rosselló, and Xiao 2010). Nevertheless, throughout the Global Finance Crisis (GFC) of 2008 and beyond, Australia managed to sustain economic stability and even growth due to the large-scale exploration of resources.

9. The FIFO workforces tend to be more cost-effective for local governments, as there is less need to invest money in community development: the families of FIFO workers are elsewhere, and therefore so are the schools, libraries and most leisure industries. Instead, FIFO workers' pay tends to support pubs and prostitutes.

10. In September 2012, the Australian owner of Hancock Prospecting Pty Ltd, Gina Rinehart, became the richest woman in the world. Her company's fortune was built on mining claims discovered in the Pilbara, including the world's largest iron ore deposit. According to the BBC, Rinehart's wealth increases by approximately $600 AUS Dollar every second.

11. Stephen Muecke can tell a story about this.

12. When in the 1970s oil was found offshore of the Netherlands, the country saw a fortune to be exploited. This led to its currency, the Dutch guilder, growing stronger. Consequently, export industries other than resources began to struggle as their products became increasingly expensive for their international clientele. Tourism is also dependent on foreign investment and is another export industry in this regard.

13. At the time of writing, the wider consequences of this decision can only be speculative. They might range from the abandonment of the *Lurujarri Heritage Trail* to a new industrial proposal for Walmadany/James Price Point and/or its adjacent region.

14. The video is accessible via the following link: https://www.youtube.com/watch?v=qIotH6GqvQI (Accessed July 17, 2017).

Chapter 5

On Common Ground

The *Lurujarri Heritage Trail* is an Indigenous tourism experience that stimulates collaboration and introduces its participants to a world different from the teleologically minded futurism of mainstream economics. It brings diverse strands of science, knowledge, law, lived experience and collaboration together, on "living country" and within a living heritage.

The *Lurujarri Heritage Trail* has further been defined as a Dreaming that the late Paddy Roe sent out and walked on—a Dreaming of "Black and White Walking Together" (Roe and Hoogland 1999). What was added to *Bugarrigarra* was this collaborative innovation, which also united the collaborators in the opposition to the LNG proposal for Walmadany/ James Price Point. Roe's vision draws on values that exceed the pursuit of economic opportunities. They are based on the recognition of what is *in place* on "living country," intra-action with(in) its living heritage, and the *liyan* this generates in people.

Earlier, I introduced "economic opportunity" as a popular argument justifying the industrialization of various places around the world and particularly remote regions. The expression is a rather ambiguous call for industrialization without much definition of in what (let alone for whom: fauna and flora, a local population, an industry with headquarters elsewhere, global shareholders?) such opportunity resides. In contrast to this, the *Lurujarri Heritage Trail* fosters a transecological relationship between people and land and as such is exemplary of alternative maneuvrings within and beyond the late liberal mainstream of Settler-colonialist Australia.

What I turn to now is the extent to which the Trail marks an Australian Dreaming beyond the resources boom. I will argue that regarding the conflict over the industrialization proposal for Walmadany/James Price Point, the Trail gives a reason for why the *Goolarabooloo* and their collaborators were anything but powerless when they were confronted by what I have termed the opportunity dreaming of government and industry. The Trail and with it the Australian Dreaming it provides access to allowed collaborators to understand their relation to the mainstream economy as "'deflated', not as their main concern, not their main business" (Muecke and Dibley 2016, 148). In this sense, it was never Indigenous and non-Indigenous protesters alone, but always both in collaboration with the *Lurujarri Heritage Trail* that successfully defeated the attempts of the resource industry and the WA government to override "living country" with large-scale industrialization, that is, "cultivated country."

At this stage, my analysis therefore returns to issues of resourceness and the different regimes of value that assembled in this conflict and in competition over the land. The following section draws on a particular ethnographic moment in which these different regimes of value—these different Dreamings, on the one hand the Trail, "living country" and living heritage, and on the other late liberalism, mainstream economics and its opportunities—collided. Whose values were more valuable? Whose Dreaming prevailed? And why?

5.1 CONFLICTING WORLD(VIEW)S AND THE POWER TO DISAGREE

> I guess we have to agree to disagree.
>
> (Kiara P., Broome Candidate for the Liberal Party in the WA Regional Elections of 2013)

In preparation for the regional elections of 2013, Kiara P., the Broome shire candidate for the Liberal Party of then WA Premier Colin Barnett, had invited tourism entrepreneurs to the Broome Tourism Office (BTO) to speak about their expectations of a newly elected shire councillor. Previously, in a personal conversation at one of her campaign booths in the Broome Boulevard, she had explained to me that for the moment—during the election campaign—she was mainly listening to people. I had asked her about her agenda, and while she had ideas, her interest was in developing these further in accordance with her electorate.

The turnout from the electorate was modest on that day at the BTO. The majority of attendees were Broome Chamber of Commerce and Industry (BCCI) members, the same institution that in 2012 had staged the Kimberley Economic Forum titled "Owning the Future," where I first met Elisabeth F. Two personalities from the Broome Airport framed the then BCCI president, Kiara P.'s husband, who had invited them on rather short notice via text message, "to give some support" (personal conversation).

This request for support added to my impression that the event had not received much attention in the wider Broome community. I counted a total of ten people, including Kiara P., the BTO director, those mentioned above, and me. Guests mingled under the influence of finger food, coffee, and orange juice. The atmosphere was polite. Kiara P. welcomed everybody and said that she would have ideas for Broome but that she had organized this event so that tourism entrepreneurs could tell her about their needs and expectations, and to discuss how she would be able to meet those if she would be elected. She was "here to listen."

Ironically, there were hardly any tourism operators present. There were no hotel owners, no Indigenous tour operators, no representatives from *Mamabulanjin* or *Goolarri Media*, the *Shinju Matsuri* Festival, the Courthouse Markets, or Willie Creek Pearl Farm, let alone the *Lurujarri Heritage Trail*. Business and industry were present, but no one who could adequately represent the unique qualities of Broome that made it a place, "nowhere else but here."

It was some time before the conversation turned to the proposed industrialization of Walmadany/James Price Point. The airport administrators saw opportunities of various sorts, including making the airport fit for an international connection to Singapore to cater for a growing FIFO workforce. An international link had long been anticipated, not least as a boost for the tourism industry. Thus far it had been deemed unviable. However, expectations of FIFO workers traveling back and forth on a regular basis to construct the LNG facility had revived the idea.

A woman I recognized from the No Gas Campaign meetings at Lotteries House intervened. She ran a bed & breakfast (B&B) in town and expressed her concerns about water and air pollution, and the impact of a male-dominated workforce and the related crowding out of restaurants, bars, and pubs on the town. Backed by the benevolent smiles of the male airport administrators, Kiara P. produced no arguments to contradict what the B&B owner said, but referred back to the "opportunities" that would outnumber what she described as "so far only speculative disadvantages" of the industrialization project.

"Broome needs to diversify its economy" was a central argument put forward by the candidate. Obviously, she had a different vision for the future of Broome from the B&B owner. She saw limits to the potential development of the tourism industry. This put her in line with the then Shire president Howard R. when he told me that "tourism won't save Broome."

But the B&B owner continued in her attempt to draw the conversation toward the qualities of Broome that were in place and deserved protection. She pointed out Aboriginal culture and heritage, the treasures of the local environment, and the thousands of recently discovered dinosaur footprints linked to the *Bugarrigarra* story of Marala/The Emu Man, all of which continued to attract people to the area. But Kiara P. stuck firmly to her point and with a disempowered gesture concluded, "I guess here we have to agree to disagree."

Back in Germany, I told this story in a workshop on conflict resolution. The "Harvard Model" (Fisher, Ury, and Patton [1981] 2011) teaches participants that the best way to solve conflicts is not to search for compromises but to return to shared interests that would reintroduce a collaborative spirit between conflicting parties. However, in a situation as described above, where the candidate of a political party sets out to listen to her electorate in order to develop her ideas for the community in accordance with their interests, agreeing to disagree is the wrong approach. It leaves no room for collaboration, but instead exposed Kiara P. as someone whose values were detached from those of a significant part of the Broome community, and who considered her values superior and did not shy away from saying so. The person who agrees to disagree considers disagreement with the other as being of no particular relevance to the fulfillment of his/her own interest.

Disagreement, in this case, was tied to a different regime of value, to a different Dreaming. Kiara P. valued industrialization and its potential benefits. She cherished the opportunity dreaming. The B&B owner valued what was in place, entangled with Broome's transecology, and was ready to preserve and protect its living heritage.

As we know now, the LNG facility at Walmadany/James Price Point was not built. "Living country" prevailed. Kiara P. lost the election, while the Green Party nearly won. Upon my return to Broome in 2017, I learnt that she had moved away. According to the internet, she took up a position as a non-executive director of a resource company in Perth.

By this time, the world(view), the values, and the Dreaming not of Kiara P. but of the B&B owner had defeated the industrialization proposal for Walmadany/James Price Point—despite the economic opportunities that were deemed likely to unfold. Without over-interpreting the fact that Kiara P. had left Broome, this adds to the impression that the Liberal candidate had not been attached to the place in the same way that collaborators in the opposition movement were attached to it.

In her campaign, Kiara P. advocated translating the values of the local environment into monetary terms, that is, the construction of an LNG facility that would override the values that were already in place. When this failed, it did not change her point of view. She still "agrees to disagree," but had to move away to pursue her opportunity dreaming elsewhere. I will elaborate on the values that prevailed over this opportunity dreaming and how an alternative Australian Dreaming stems from their lived experience.

5.2 AN AUSTRALIAN DREAMING

Like everything else, the *Lurujarri Heritage Trail* and with it the *Goolarabooloo* can be positioned within the wider monetized and media-saturated world as we know it, with related (welfare) services supplied by the state. However, the Trail is different because it has never sought or received any promotional or organizational support from government funding bodies. A rare exception was the previously mentioned donation of a used truck by the late Lord Alistair McAlpine to Paddy Roe in the very early days of the endeavor.

The Trail is self-sustaining on various levels, including its finances. Promotion has occurred solely via word of mouth and more recently via social media. As of 2015, costs are covered by the participation fee of $1,600 AUS Dollar per person for the nine-day walk, with concessions for children and students. All of the income generated flows back to the *Goolarabooloo* families and is used for the upkeep of campsites and the payment of bills. It covers expenses for food, fuel, a truck, 4WDs, and the like that are used to transport the gear of Trail participants (and sometimes the exhausted participants themselves), as well as to cater for the *Goolarabooloo* family members that accompany them (figure 5.1).

This independence from external funding regimes provides further evidence of why the *Lurujarri Heritage Trail* is positioned outside of interpretations of it as small or marginal within the mainstream economy. The transmission of the values of the local environment, of Indigenous heritage and culture, has arguably been of the same if not greater concern for the *Goolarabooloo* than a translation of those into monetary form. The emphasis on teaching and learning about an ontologically different world(view), and of making this world(view) accessible to others through lived experience, is a unique quality of the Trail, as to preserve "living country" meant both engaging with the Settler-colonialist situation and offering a means of doing so in transecological terms in order to stay true to its Dreaming.

A comparison with other Indigenous tourism ventures highlights this crucial characteristic of the Trail. Most advertise the experience of unique fauna and flora through the purchase of an environmentally friendly Indigenous tourism product. Standard promotional material available at local visitor centers in the Kimberley and elsewhere predominantly

Figure 5.1. Packing Up and Loading the Truck with Swags of Trail Participants. Photo: Carsten Wergin.

emphasizes remoteness. The "authentic" Indigenous tourism experience is generally considered to be remote in time as well as space (Lane and Waitt 2001, 387). Tour guides highlight the unique natural environment and suggest activities for independent travelers while continuing to "refer to Aboriginal culture in terms of rock art sites that signpost their journey in a timeless land" (ibid., 393).

The *Lurujarri Heritage Trail* is marked out as different from such conventional Indigenous tourism products because its very experience is dependent on its participants' active engagement in it. In order to stay true to its Dreaming, participants are required to strip bare of preconditioned assumptions and aim to accept the lived experience "as it is." Conversely, the trail also relies on what its Indigenous and non-Indigenous participants do on "living country" in the course of this experience. As such, the Trail experience is anything but remote. It rather becomes an existential part of the very life-journeys of its participants which literally entangle within its transecology on various aesthetic levels such as story, scent, and sweat.

Despite the conventional tourism industry's continued disregard for the hidden qualities of this Indigenous tourism experience, such experiences readily suit the interests of many travelers who enjoy camping, fishing, or hiking trips (Buzard 1993; Cohen 1979). Linked to the direct engagement of participants and a free-floating organizational structure, the Trail therefore leads beyond conventional celebrations of Aboriginal culture as "frozen in time" and as represented in art, music, or dance performance. Instead, the collaborative experience of walking across "living country" and learning about its living heritage transmits values that offer a very different point of view on the environment, whose well-being becomes entangled with the well-being of everyone else.

At the end of their journey, the collaborative experience of walking the land, which is accompanied by activities such as fishing, collecting bush honey, observing the ocean, the stars, and the large tidal movements, enables participants in the Trail to feel attached to the land in a similar way to their hosts. In this regard, the experience of "living country" becomes a mediator between different ontologies that generated significant support when it came to preserving it from industrialization. The argument advocated by *Goolarabooloo* was that the construction of the LNG facility on such "living country" would sever their entanglement with the environment. Therefore, not only would the *Lurujarri Heritage Trail* be destroyed, but their spirits would suffer. The industrialization project could result in cultural genocide. This threat extended to Trail participants, who in the course of the nine-day walk learn to acknowledge "living country" as the common ground from which to overcome the dualist understanding of Indigenous and non-Indigenous world(view)s that sees one as superior to the other.

Many who have experienced the Trail in this collaborative sense have been students of arts and landscape architecture from the Royal Melbourne Institute of Technology (RMIT). The collaboration between the *Goolarabooloo* and RMIT began in the early 1990s and can be deemed central to the establishment of the Trail. At the time, Jim Sinatra, a professor of U.S. origin, took up a position as Head of Department of Landscape Architecture at RMIT. He was interested in Indigenous perceptions of the environment and ways in which to teach these concepts to his students. Jeanné Brown, whom I introduced above as one of the long-term participants and intermittent co-organizers of the Trail, explained to me how Sinatra had worked with American First Nations peoples and therefore understood the importance of having some appreciation for the cultural connection that Indigenous Australians feel to the land. However, on arriving at RMIT he found that this was not part of the syllabus of landscape architects or any other students for that matter. Sinatra sought a means to change this and found the *Lurujarri Heritage Trail.*

In 1992, Brown was a friend of one of the students who first accompanied Sinatra to Broome. While living up the coast on the Dampier Peninsula, this friend sent her "great descriptive letters of the cultural life here;

the wildlife and the fishing and the language and customs" (Transcript 121109_JB). At the time, mining companies were exploring the mineral sands north of Walmadany/James Price Point, and Paddy Roe sought Jim Sinatra's help to document the names and traditional stories of places and plants of significance in Aboriginal Lore and culture. As pointed out in previous chapters, from early on Roe recognized collaboration with non-Indigenous people as a vital means of keeping "living country" and its living heritage alive. This included the documentation of coastal vegetation communities and functional plant use along the Song Cycle.

In one of his letters, Brown's friend asked her to volunteer for this documentation project. She had drawing skills and prior knowledge from work in plant nurseries that included "Latin name exposure," as she told me with a smile during our interview. We sat on the balcony of our apartment on Dora St, with views across to the Old Hostel from which Trail participants set out each year. Brown pointed to the old tamarind tree under which Paddy Roe would sit and discuss the heritage and culture of the region with Sinatra and his students (figure 5.2).

In 1992, the year of the Mabo decision, cultural documentation became the focus of a Final Year RMIT project led by Sinatra. He and his students gathered information. They mapped the country along the *Lurujarri Heritage Trail* between *Minyirr* and *Bindijangun* during extended bush forays,

Figure 5.2. Trailers at the Old Hostel under the Tamarind Tree. Photo: Carsten Wergin.

and spent hours with Paddy Roe under the tamarind tree, discussing and listening to stories, often while learning the skills of artifact carving. The group was based at *Millibinyarri*/Coconut Wells, which at the time was "just bush, and only Frans [Hoogland] was living there" (Transcript 121109_JB). Brown explained why Frans Hoogland became central to the *Goolarabooloo* community at the time:

> He's got the English. Well, he's got sort of Dutchman's English [laughs] but compared to Paddy's English skills Frans was more conversant with basic legal terminology. They [*Goolarabooloo*] call it "white magic": understand the conceptual framework of Western civilisation, the legal courts. (Transcript 121109_JB)

Brown confirms that from early on, Paddy Roe recognized the importance of "white magic" and therefore the need to collaborate with those conversant in it to make his concerns heard. For this reason, together with Frans Hoogland and his grandsons, mainly Law Man Richard Hunter, he began to take groups of *kartiyas* out bush to teach them about 'living country.' In turn, his hope was that their documentation of this on paper, translated into Western scientific terminology with white magic," would later protect the coastline and the Aboriginal Lore and culture it holds from looming mining interests.

The Trail was inaugurated and became a further means to teach about Aboriginal Lore and culture "on country." The bonds that form between Trail participants—some of them RMIT students, others casual tourists, and travelers—and "living country" continue the collaborations that Roe began with people like Sinatra. Some former participants returned to protect the Trail from industrial development, for example a marine biologist and his partner, who contributed their skills in "white magic" motivated by what they had learnt to appreciate about country through its lived experience with *Goolarabooloo* family members.

In 2012, I walked the *Lurujarri Heritage Trail* myself for the first time and conducted interviews with other first-time participants. While it was certainly different from when Jeanné Brown went "out bush" in the early 1990s, some parallels remained. For one, a group of participants still was made up of RMIT students with an interest in learning more about Indigenous Australia. As an Australian-born RMIT student in his early twenties explained, "We get to choose [a course] and it really hit me because I've wanted to come up north to experience Indigenous culture, 'cause I think it's very important for Australia. But also to learn more about the diverse landscape up here. And it is so beautiful out here. The landscape just keeps changing. It's really wonderful" (Transcript 120707_anonymous).

For sixty years and more, Australian anthropology has shown how "the relations between a group and its territory are among other things ecological relations" (Stanner 1965, 1). The *Lurujarri Heritage Trail* and the way it is run by the *Goolarabooloo* manifests a space in which such teachings can be experienced. Within the diverse group of people that participates in it, conventional hierarchical relations between them and also between them and the landscape they traverse are renegotiated (see Povinelli 1995, 510–11, for a similar discussion).

With Povinelli, I argue that fundamental to this is what she calls "labor" that is invested by all participants in the landscape: "People who live and labor in an area permeate the region with their words and sweat" (Povinelli 1994, 152). Meanwhile, the labor of the *Goolarabooloo* has remained a central point of conflict between them and other Indigenous groups. When Woodside withdrew from building the LNG facility, the Trail remained as the foundation for longer-lasting collaborations between people and place. So did a $30 million AUS Dollar compensation package. As I discuss in the following chapter, with this money as "the elephant in the room," the conflict over the LNG facility continued in front of a Native Title tribunal. Now that Native Title has been given to the Indigenous group that supported the industrialization project, the survival of "living country," and with it the survival of the Trail, is again under jeopardy (figure 5.3).

This offers a devastating scenario with regard to the future of the *Goolarabooloo* community. However, the continuous labor of the *Goolarabooloo* on

Figure 5.3. Trail Participants at Walmadany/James Price Point. Photo: Carsten Wergin.

country has created significantly more than mere respect for them within and beyond the Broome region. For one, it has empowered them to stage arguably the most substantial protest action against looming mining interests in Australian history. In addition to this, their labor has amalgamated them with "living country" no matter whether a court officially recognized them as its custodians or not.

Difference lies in the fact that proponents of "economic opportunity" emphasized that the land around Walmadany/James Price Point needed an industrialization project in order to become valuable, while the *Goolarabooloo* and their collaborators identified its value in the here and now. The Native Title decision recognizes this value in the here and now but based on legislation grounded in Western law and culture. This is tragic for the Trail, because its transecological qualities, and along with those its collaborative values, are not recognized. Neither the labor of the *Goolarabooloo* nor that of the annual Trail participants who have also walked the land and connected with it through "their words and sweat" (Povinelli 1994, 152) are relevant in this scenario.

Such amalgamation is the potential source of an Australian Dreaming post-resources boom. The remainder of this chapter will flesh out the resourceness of this Dreaming, the particular values it carries, and how those relate to the qualities of the Trail. As Roe and Hoogland put it, the Trail might be forced to return into the ground—overshadowed by "cultivated country" and "hiding, withdrawn, waiting." But its Dreaming prevails and will see the Trail re-emerge as part of a world in transecology.

5.3 TRANSECOLOGY IN ACTION

On the Trail, things tend to simply happen, and one is either part of them when they do or one is not. As a result, and as I witnessed one evening around the campfire, some students approached their teachers to exclaim, "I'm not learning anything!" During my third participation in the *Lurujarri Heritage Trail* in August 2015, I had a conversation with two of the new generation of *Goolarabooloo* about the potential knowledge transmitted to other Trail participants once they confront and overcome such concerns. Jonathan D. and Mick H. were both in their twenties to early thirties at the time, and gradually to take over from Richard Hunter and other Elders, for whom the work as guides, storytellers, and singers had become too physically exhausting:

> *Jonathan D.:* I just love walking country and meeting different people. I prefer to be with the older guys (. . .) They are in their 60s, 70s [and] probably grew up in that environment where they only had word-of-mouth about how Aboriginal people are living. Most Aboriginal people are portrayed as

> being drunks and alcoholics (. . .) And just to change that view . . . how we got respect for country and walk through country and still live our culture and respect our Lore. You have to change their minds. So when they sit around their family, talk about Aboriginal people, [they] put them in a better light . . . when they go back home to wherever they come from and to their grandkids and their kids. (Transcript 150817_TH)
>
> *Mick H.:* Trying to look after the country and meet more tourists and make more tourists to come back to our country. . . . Also, the country brings them back sometimes. Their *liyan*. . . . That means feeling. They feel it. This country makes them happier. And they like the ceremony, what we do, dancing. I like to talk with them . . . listen to them and how they talk about where they come from . . . and spin out how their culture is different to us. They say we've got better culture, better lifestyle, culture-wise, then they. Because their culture got cut out a long time ago (. . .) Their culture died . . . lost (. . .) The old people didn't teach the young generations to carry it on, like how our great-great-grandfather Paddy Roe showed his grandsons. (Transcript 150817_JJ)

Both Mick H. and Jonathan D. emphasize how both Trail participants and they themselves learn through exchanges with each other and experiences related to the land. This includes culture and family ties, Aboriginal life, and relationships to country. Both describe how their exchanges are mutually beneficial and as such collaborative. However, these exchanges also remain challenging, because their loose organization provides only possibilities for what might happen prior to the actual experience. As Jeanné Brown explained when we sat on the balcony overlooking Dora St:

> You think about the *whitefella* culture, it's very much obsessed with structure and predictability. But people tend to do things here when the time feels right. We stay at set camps on particular days but beyond that it's a much more organic process—if people go crabbing, or if they go looking for bush honey. There is a lot more flexibility in terms of what the plan might be for any one day. And it also depends on what the weather is doing, or things that happen! Things you discover. You don't want to cut yourself off from being able to pursue something that happens on the day. Like last year, when we were up at *Bindinyanygun* / Yellow River: Richard [Hunter] took a walk up Hidden Creek and they (. . .) tracked a goanna that they dug up and cooked (. . .) So it is constantly new things that the country presents to you (. . .) But it's also about it not being a fitness race. It has a more experiential dimension of being in a place.
>
> The first day or two people are still talking about the world they have come from (. . .) It takes a while to actually be in the moment and be in the day that you're in and not be focused on abstract stuff from elsewhere (. . .) Or even learn that it's ok just to walk and be quiet; that you don't have to fill every moment with talk; that it's ok to travel quietly, take things in (. . .) Give yourself a chance to become aware on other levels. That's not an easy thing for a lot of people. It takes a while to change modes. (Transcript 121109_JB)

In Jeanné Brown's, Jonathan D.'s, and Mick H.'s accounts, the *Lurujarri Heritage Trail* has great value for its participants because it challenges them to encounter world(view)s other than, for example, those that make up student life in Melbourne. The Trail demands that they leave "abstract stuff" behind. This also includes stereotypes about Aboriginal people that they might have become accustomed to. Instead, people "travel quietly, take things in."

The Trail is an "organic process" in this regard, as Brown terms it. It entangles people with the environment in both a spiritual and a physical sense. It allows for those participating in the Trail to immerse themselves in the landscape—to become part of its transecology. This entanglement of participants with the environment even alters the experience the Trail has to offer. Personal engagement with it, the "labor" performed on "living country," is what makes it a living heritage that is of existential value for both people and the environment.

In further explaining this point, I return to the motivations of both Brown and the RMIT students. Brown, with a background in drawing, education, and nature conservation, had a particular interest in the diverse ecosystems encountered by participants:

> You start on open beach, Cable Beach, and dunes. Then you cross salt marsh flats, through mangroves and creeks, and through the back of the dunes into wetlands. Then through the mouth of Barred Creek mangrove and down to the reef. Or you go crabbing in the low tide mangroves and you get to walk right up the fingers of the creeks. It's really diverse: The different habitats, the different types of country. And they are not all separate. They all link together. And the transitions between one place and another: let's say you camp at Barred Creek, and you see it across the period of a day. You might have an extremely low tide in the morning, but then you see the tide come in and the activities change depending on how much water is in the body of the creek. Where you fish from. Or, when the tide is out you can go crabbing. Spear a stingray. Explore the mangroves. And the whole tidal differences: you wouldn't even recognize it as the same place because the tidal reaches are so extreme. They are eight, nine-meter tides. . . . And the country that gets exposed as the water sucks in and out of it, over six hours—it's quite amazing. (Transcript 121109_JB)

As Brown explains, the land, the ocean, the moon, and the tidal movements align in the transecological experience the Trail offers. She ties this back to childhood memories:

> As a kid, my family used to spend time every holidays camping down at the coast at Sorrento, which is on the Mornington peninsula, in Victoria (. . .) Broome is very similar. Different colours, but equally wild and reefy, and lots of rock pools, and amazing marine life, and huge tides, rips, tidal pools.

> At the time, there were big middens along the top of the cliffs. But as a kid, I didn't understand that they were eating grounds. I couldn't understand how the sea had got up that high to dump all these shells on the cliffs. They never taught me anything in school about the Aboriginal presence in that area (. . .) These were their old camps, and a lot of the same shells occur there as here [Figures 7.5-7.8]. But I remember as a child being quite mystified by it. I couldn't get my head around it. It was not until I came to Broome and was taken bush and camping in all those places up here that it clicked with me that (. . .) there had been people living in the places that I was most connected to in Victoria. (Transcript 121109_JB)

The trail is a collaborative experience that criss-crosses space and time. Its transecology offers participants a different perspective on the land they are traversing, as well as new insights into their personal histories and childhood experiences. It allows people to imagine the latter as not separate from but entangled with a wider Australia, which as will be seen below has been a longing for many non-Indigenous people growing up in a Settler-colonialist society with racist, white supremacist, segregatory principles that they loathe and seek to overcome.

Collaborations with Indigenous guides and the engagement with "living country" have a profound impact on Trail participants as they resonate with individual life histories and transform those. This experience is fundamentally different from conventional tourist experiences aimed at providing economic opportunities for Indigenous hosts, because Trail participants are more than mere guests. They turn the Trail into part of their own journey, rather than buying into pre-conceived attractions the remnants of which are postcards and souvenirs, and this journey marks a shift in their perception of what Australia is and can be. Mick H. recalls an experience like this that he shared with a woman from New Zealand in her mid-70s: "I stayed back with her and walked behind. She said, "I don't know what they're missing in this place. Because it's very beautiful country and it's best to walk a bit slowly. At least you can see everything. See the tracks and whatever . . . on the sand" (Transcript 150817_JJ). The example highlights the aesthetic and sensual dimensions of the experience, which are not representable in Modernist terms, and the collaborative relationships between people and land that these foster. This adds to why "white magic" cannot properly account for the Trail experience, which is fundamentally different from the opportunity dreaming that has driven Multicultural Australia.

Again, providing an experience that allows participants to "step back" and "take things slowly" has grown out of Indigenous traditions as much as out of necessity. In the early 1990s, the *Goolarabooloo* and their understanding of country were marginalized in mainstream economic terms, and endangered by continued exploration initiatives. Furthermore, as outlined above, the kinship, culture and family histories of all Indigenous people were under significant pressure prior to the 1992 Mabo decision. They were meant to assimilate into the mainstream political and economic system, or in other words, to leave their myth-consciousness behind and inscribe themselves and their culture into the Settler-colonialist world(view).

In response, Paddy Roe for one invited many people to help document the significance of "living country" and living heritage. And he had further concerns, as the Trail was also meant to bring his own extended family back on country, as Brown explains, "When I first came here, all his family were living in town (. . .) He [Paddy Roe] wanted them to be out bush more often, in relationship with the country and their cultural practices. You easily get seduced by the television, and the refrigerator, and the basketball, and videos, and all those sort of entertainment dimensions of the town" (Transcript 121109_JB).

Mick H. adds to this an explanation of how Paddy Roe's grandson, Mr Roe, continued the *Lurujarri Heritage Trail* as a means to bring a new generation of *Goolarabooloo* back on country:

> I was living in Derby with my father and mother and just when I got into high school, I ended up staying with [Mr Roe]. He took us out camping and showed us things about country. How to make fire and look around, do everything. And then [he] said, "this is what you're going to do when you get a bit older. You've got something to do on the Trail then." And when school holidays came we'd go on Trail (. . .) Smile, walk with the tourists . . . have a good time with them. Show them our little areas where to swim and things (. . .) teach them how to throw a line in. (Transcript 150817_JJ)

And Jonathan D. explains how the Trail unfolded for him in this regard:

> We [him and Mick H.] have always been sticking together since we were very young so we'll carry that thing on. We don't really need other brothers but it's good to have them here with us [to help with] some of the stories that I haven't got fully and other brothers picked up on real quick. It's good to have them there as well if I can't get that story properly always. Give them the elbow, "your turn to talk now." (Transcript 150817_TH)

As Paddy Roe was well aware of the immediate threat of the *Goolarabooloo* losing their identity and tradition, engagement with the Trail was to guarantee that the extended *Goolarabooloo* families continued to nurture their connection to the land. Walking the Trail has by now become a collaborative experience for every generation of the *Goolarabooloo* that underpins their ties to "living country" and a wider community of like-minded people as much as to their own family history and position within the region. Paddy Roe's Dreaming became manifest in the *Lurujarri Heritage*

Trail, through documentation, the consistent presence of his family on country, "labor," and the shared experience of its living heritage. This protected the land against the construction of the LNG facility, against the opportunity dreaming of government and industry. And it forms the basis for an Australian Dreaming that goes beyond opportunity.

5.4 DREAMINGS BEYOND OPPORTUNITY

The practical side of Paddy Roe's Dreaming can be described as twofold. First, through documentation, he aimed to have control of how the country he felt responsible for was understood and spoken about, and second, by establishing the Trail, he sought to control what was done on it and who would function as its future custodians. Jonathan D. and Mick H. emphasize that the goals of their great-grandfather (*Lulu*) had been achieved, at least up until the Native Title determination turned against the *Goolarabooloo* and in favor of the *Bindunbur* claimants. However, while Native Title might have taken the land from them, it cannot override the Dreaming that manifests in its transecology, not least because each time industrialization proposals have been put forward they have strengthened rather than weakened the Trail and with it the *Goolarabooloo*. As Brown pointed out in 2012, when protests against the LNG development had reached another peak:

> Over the years, different family members have become more involved in organising the Trail. It's never exactly the same team of people that run it. It depends on who is available, especially now with all the extra demands on all the family's time, with this pressure of the industry looming, and having to deal with lawyers (. . .) You know, last year [2011], right in the middle of the Trail was "Black Tuesday," when all the machinery was to make its way back into country to start clearing it. So most of the family had to be present at the Manari Rd blockade (. . .) Thankfully, the Hunter family came and helped carry it on. That was a great relief (. . .)
>
> On that Black Tuesday we were at Barred Creek and that morning we got the word that this was likely to happen and so we set up the breakfast—I had about ten volunteers helping me—and we all went off to the blockade in the morning. That's when all the conflict happened, 40 cops and riot police and that. The Trail people were walking from Barred Creek to Quondong that day and I knew I had to be back to do the cooking. But at lunchtime this major conflict was all still going on at Manari Rd corner. I knew I had to leave, and few of the volunteers would come back! They all wanted to stay and I was left with nobody to help. But all the students kicked in (. . .) I was really stressed. Because I had wanted to stay at Manari Rd, too. But I knew the most important thing was to keep the Trail going and celebrate

> the positive stuff that we had set up. We couldn't let that fail, because that was the whole manifestation of the vision of [Paddy Roe] (. . .) The students helped me until my volunteers came back, about three days later (. . .) I was very stressed [laughs]. But it ended up being a fantastic Trail! And thank God all the Hunter family were there. 'Cause normally they are more in the background; do lots of their fishing and that bit. They don't talk so much to people, because they are a bit more quiet and shy. Yeah, it was very interesting. (Transcript 121109_JB)

Brown highlights the diversity of the *Goolarabooloo* family members who commit themselves to the upkeep of the Trail. Then she explains how Trail participants quickly become attached to the land and involve themselves in its protection. After the interview, I came to ask myself, where would I have been? Would I have remained with the Trail, or would I have left to experience Black Tuesday, maybe side with the *Goolarabooloo* and those blocking Manari Rd?

The choice between these two options is less relevant than the fact that Trail participants played a role in both, as protesters at Manari Rd and as helpers to keep the Trail running. By this point, a day or two after Willie Creek Pearl Farm, after leaving "abstract stuff" behind, the Trail experience had stimulated in participants the means to knit the Indigenous culture, its "living country" and living heritage, into their own life stories. Protecting the Trail became existential to them. This shows clearly how the Trail experience nurtures transecological collaboration, which in turn preserves the land. As a result, in 2011, during the events of Black Tuesday, even first-time participants took action against government and industry.

Such collaborations lead well beyond conventional dualisms, as they acknowledge ontological differences and diverse world(view)s as equal to each other. They take place not only between Indigenous and non-Indigenous people but also between people and the Trail itself. In this way, the Trail has been able to reconcile an Australian people, something that generations of Indigenous and non-Indigenous politicians, government, and policy-makers have failed to do.

The anthropologist Jon Altman criticized diverse projects run by successive Australian governments with the aim of assimilating the Indigenous population into neoliberal ideology (Altman 2010, 262). Against an alleged need for "assimilation," "intervention," or "closing the gap," to paraphrase some of the more prominent names given to these projects, Altman argued that Indigenous people have developed their own ways of incorporating the late-liberal mainstream of Setter-colonialism into their everyday lives that are not discernible by way of conventional statistics about wealth and well-being. These alternative means are characterized by a "post-colonial adaptation (. . .) that sees engagement in all sectors of the local economy" (Altman 2007, 4).

According to Altman, such engagement can be traditional (fishing), market-oriented (tourism businesses), or state-led (receiving social benefits). While it is widely acknowledged that any market economy can only function if it is "embedded" in a societal system that caters for it (cf. Karpik 2010, 111; also Polanyi 1944), the resulting "hybrid economy" of Aboriginal communities has been largely overlooked. Australian policymakers focused instead on the spread of a Western market model—or in terms of the argument made here, they focused on the development of opportunities, rather than fostering a transecology engagement that was already in place.

Food supply during the *Lurujarri Heritage Trail* is another example of this hybrid economy. In figure 5.4, Trail participants help prepare dinner, while the meat of a recently shot bullock (locally known as a "killer") dangles on ropes fixed to a roof structure. The animal had been hunted that same day by *Goolarabooloo* family members and transported back to camp in a 4WD Toyota. At the camp, it was skinned and its meat prepared for sharing amongst the group. Fresh Kimberley beef in the Australian outback: a delicacy and an expensive one in mainstream economic terms.

Preparing and eating a freshly hunted "killer" was part of lived experience, the system the Trail is both embedded in and expression of. It was also part of the "labor" performed by those immersed in this "living

Figure 5.4. Skinning the Killer. Photo: Carsten Wergin.

country," and thus stands out from the mainstream economic system and its conventional tourist experiences in a hybrid way. As "local circumstances [. . .] require local solutions; solutions that [are] not just [. . .] driven from outside (or top-down) and that [. . .] need to mesh with local aspirations" (Altman 2004, 530). Rather than shopping for meat in a supermarket, Trail participants were engaged in a sensual experience that connected them to the land in an unprecedented way through a further means of "labor," alongside walking, sweating, fishing, and setting up camp.

This labor, "living country" and its living heritage—stories as much as the hands-on experience of skinning a 'killer' instead of buying a Scotch fillet in the supermarket—allows for participants to conceivably become part of the community and the environment themselves. *Goolarabooloo* hybrids? Collaborative economists? Jeanné Brown might describe herself as just that. In a personal conversation at the end of the 2012 *Lurujarri Heritage Trail*, another participant added:

> *Trail Participant:* I guess, this experience is very real [laughs and points at the "killer"]. Because, in any other tourist trip we would not see any of this going on.
>
> *Carsten Wergin:* Did you see the big turtles, too? The ones they caught?
>
> *TP:* No, I did not see them. I did taste a bit of turtle (. . .) I made some clapping sticks . . . I should say that Richard [Hunter] did a lot of it! [laughs] We only had the one good axe. And the one I was trying to use was not the good one. Nothing was moving, so Richard just took it and he just did it. So I just sanded it down.
>
> *CW:* Richard felt pity . . .
>
> *TP:* [Laughs] Yeah, he helped us all with it. There was a few of us making clapping sticks and he helped us. It was really great! [Pause] Now I might go fishing. (Transcript 120712_anonymous)

The crucial point made by this participant is that for her the *Lurujarri Heritage Trail* and its collaborations became "very real." Plenty more adjectives could have come to mind while observing the "killer" being skinned: "brutal," "strange," "foreign," "different," to say the least. But for her it was "real," because she was not simply observing it but participating in it. The Trail participant was part of the intra-action, the performance of reality, of transecological entanglement with the land.

In this regard, our brief exchange ended on another significant note: her spontaneous suggestion that she might go fishing in the Indian Ocean, something she would not have done or even have had the knowledge to do before she joined the *Lurujarri Heritage Trail* and as a result became part of "living country" and living heritage. Was she starting her own hybrid economy? Maybe not. Her intra-action, however, highlights how her participation in the Trail and the related labor she invested lead beyond Settler-colonialist visions of Australia in that they stimulate a transecological Dreaming of "Black and White," and also of human and more-than-human, walking together on country.

5.5 HERITAGE-MAKING ACROSS MULTIPLE WORLDS

The attachment, collaboration, and hybrid experience of being in the world that these and the following ethnographic examples illustrate, whether in the forms of heritage documentation, citizen science, or Trail participation, will present diverse generations of Indigenous and non-Indigenous collaborators from a multitude of backgrounds who are united in that they have never fully acknowledged the teleologically minded futurism that motivates striving for opportunities. Their stories are rather evidence of a world(view) beyond the "artificialism at large" that Louis Dumont (1986) criticizes, or as I term it, an Australian Dreaming after the resources boom.

In 2015, Trail participants (TPs) shared some of their own related experiences on Trail with me when they met for farewell drinks at Matso's Restaurant in Broome. I joined the group and asked randomly what they took away from their participation in the Trail. Below are almost unedited excerpts from the first ten minutes of my conversation with one male and one female participant, both Australian born and not from the Kimberley region. I engage with those in more detail because the stories they shared resonate with many central issues that have come up in the course of this inquiry.

> *Trail Participant 1:* In the middle of that hike I was like, "This is weird." Where is the information? When are we going to get all the stories? We are not learning anything . . . I was really coming with that Western mindset. #00:00:44-9# (. . .) Now at the end, last night, it really came together. We were learning about the energy of the place. We are learning about being quiet. Everybody had slowed down. When we were talking we were looking at each other (. . .) People had learnt to listen to each other. #00:01:20-4# (. . .) The most beautiful for me was when Aubrey was singing. And you could feel the pain that through the centuries had been inflicted on their culture (. . .) I could feel it not intellectually but emotionally.
>
> *Carsten Wergin:* You mean Aubrey or Phil? (. . .)
>
> *TP1:* Aubrey was playing this Irish song, and there was this emotion that came through. #00:02:04-8# (Transcript 150823_anonymous)

TP1 set out from his expectations, his frustrations at not learning anything in the first days, and how these transformed into an immanent transecological experience. Representationisms were left behind, as silences and regard for the other became more important to him than the fulfillment of the expectations he had brought to the Trail. Emotional not intellectual stimuli became relevant. Those included the energy and *liyan* that came from the land as much as the music that he encountered during the *Corroboree* on the final night.

Interestingly, what TP1 remembered most in this regard was not Phil—who as always at the end of each Trail performed Indigenous songs that relate to the country—singing, but the Irish song sung by Aubrey, a Melbourne-born musician and excellent blues guitarist with whom I had the pleasure of playing saxophone myself, who was invited by fellow music-friends to join the *Goolarabooloo* in their opposition to the industrialization project and has been with them ever since.

My conversation at Matso's continued with an immediate intervention by Trail Participant 2:

> I felt the same way. I was dying . . . I was hungry for information (. . .) Then I realized, it is not really information I was looking for. It is really my connection to the land . . . through its people. I feel now more equipped with the answer, when somebody says, "Where are you from?" #00:03:16-8# (. . .) When people asked me I'd say, "I'm from Australia, 200 years old, Opera House, Bridge . . ." Now I say, "I'm part of this culture, which goes back so many thousand years" (. . .) So I feel Australian because I'm connected to its people. #00:03:51-8# (. . .) I feel satisfied. I'm happy that I went through it. And I guess the friendship we created, whether its shorter or longer, this is what it's all about. #00:05:03-5# (. . .) In regards to learning, I'm dying to go back and find out more in regards to what's happened here (. . .) Tess [a volunteer helper on Trail] was talking about land rights. . . . It made me feel how selfish we are. Because, when I hear about it I don't care. But now I feel [that], if I hear that something could happen, I'd get there and walk with them. #00:05:39-1# (Transcript 150823_anonymous)

When TP2 set out to participate in the Trail, she was as "hungry for information" as TP1. After the Trail experience, she highlights the relationships that formed, the relationships she established with people and place, and how those will impact on her personal life story from now into the future. Her exploration of the country as a participant in the Trail allowed her to think about and present her Australian identity in a different way from before, as part of a culture that goes back thousands of years instead of only the couple of hundred since the arrival of Captain Cook.

The Trail experience allowed for TP2 to no longer separate Indigenous and a non-Indigenous Australian history because of a now-shared commitment to the land. In the future she is prepared to defend this new Australian Dreaming she found, or as she put it, "If I hear that something could happen, I'd get there and walk with them."

Our conversation then turned to individual motivations for signing up to participate in the Trail:

> *TP1:* I came because of the gas plant. Because that was the first time I had heard about the Lurujarri Trail. And on the photos you don't see how beautiful that land actually is. There is no photo that could ever do it justice. To sit there and look at the rocks at sunset, you go, "How could anyone in this world ever think of destroying this?" #00:09:00-0#
>
> *TP2:* I went to this place called Port Macquarie, north of Sydney. They are selling African artifacts. And I got to talk with the lady at the cashier about walks. I said, "I like walks." And she said, "You should do this walk!" And I said, "OK." So I booked. #00:10:52-2# (Transcript 150823_anonymous)

Noteworthy in both cases are the unconventional ways in which the two participants came to know about the Indigenous tourism experience that is the Trail. TP1 was attracted by media coverage about the gas plant. He was eager to get to know the place that motivated so much protest. Eventually, he made a similarly heartfelt connection to it that left him speechless as to how it could ever be up for industrialization.

TP2 learnt about it through word of mouth, which is an indicator of how the experience of the Trail and with it the Australian Dreaming extends across the country, in this case into a shop for African artisans in Port Macquarie, near Sydney. Through such unconventional means of distribution, the Trail attracts like-minded people to its transecological engagement and possibly more fundamental change of mind.

These final examples show that through the *Lurujarri Heritage Trail*, a transecological alternative to the Settler-colonialist mainstream has been in place for over thirty years. Instead of aiming for mastery over the environment, the Trail and its participants collaborate with it. Meanwhile, market economics saw no opportunities in such collaboration, and consequently ignored the Trail and its Dreaming. This ignorance and arrogance of the Settler-colonialists, which makes them assume that they have the power to "agree to disagree," remains the most fundamental obstacle to reconciliation—or, as Jeanné Brown pointed out to me in frustration:

> These [Aboriginal] people knew their country so well but nobody was bothered asking them about how to do anything. They had all this expertise, incredible skills at fishing, knowing what's edible, finding water, finding honey, tracking things, looking out for snakes, but nobody, none of the Western culture that had come to live in Broome, nobody was actually consulting them or even seeking their help, or seeking their opinion on how best to live here. (Transcript 121109_JB)

Meanwhile, the diverse collaborations I have presented above that have emerged from the *Lurujarri Heritage Trail* prove that there are places that offer ways of reengaging with Australia's living country and living heritage through collaborative labor. Such heritage is a messy field of transecological production. It is made up of lived experiences of what the responsibility to keep country and culture alive entails. But this Dreaming also gives people the *liyan* to cultivate transecological collaboration, something that in times of global crisis appears almost too radical to hope for.

Chapter 6

Knowledge and Place-Making

Diverse materialities such as grinding stones, turtle bones, and fireplaces that participants find while following the Trail invoke a living heritage that is continuously reworked by both custodians of the land and more-than-human actors like wind and rain. What Indigenous people ironically term "white magic" must be applied to these intra-actions in order for them to be representable in courts of law and other Modernist institutions that seek to determine who holds Native Title over a particular stretch of land, or "country." My personal involvement in the process of applying such "white magic" is discussed later on (chapter 7).

For Native Title law to be applicable, decision-makers require evidence of continuous cultural practice on country from applicants. Such an expectation falls short of recognizing the intrinsic values of the land—the processual dimension of a transecology that is based on multispecies entanglement. In the eyes of the law, cultural practices only happen "on country" in the sense that country provides a place where they are performed. In Indigenous terms, however, the country itself is what gives life to cultural practices. It is not the stage for, but rather an actor in, the performance of Aboriginal reality.

While the agency of "living country" is downplayed or ignored by "white magic" and other Modernist constraints, it had great relevance to the successful opposition movement against the LNG facility. This chapter makes this relevance evident for the sphere of collaborative knowledge production. I draw on material collected during (1) the final days of the *Bindunbur* Native Title hearing in April 2017, and (2) citizen science projects conducted between 2011 and 2013 as part of the Walmadany/James Price Point protest actions.

The Native Title hearing highlights another wicked problem embedded in the attempt to reconcile Aboriginal Lore and culture with Western law and Settler-colonialism. This problem is linked to ontologically different perspectives on the land, its heritage values, its resourceness, and what is implied if one claims to *own* it. Native Title determinations function by attributing ownership of the land to Native Title holders. But what happens if an applicant does not base their claim on ownership of the land but on responsibilities for it, as did the *Goolarabooloo*?

Following on from my discussion of ownership in regard to the *Bindunbur* Native Title case, the citizen science projects will highlight further ways in which the wider Broome community was able to not only actively engage in the conflict over Walmadany/James Price Point but also significantly influence its outcome through knowledge production. Those projects were set up in response to the social and environmental assessments run by state and industry consultant firms. The latter presented results in ways tailored to the needs of the proponents of the project. The potential environmental effects of the LNG facility were acknowledged but framed as inevitable variables within more far-reaching calculations that emphasized the economic "opportunities" at hand. The benefits from profit maximization through resource extraction added up in favor of the project going ahead, despite environmental losses.

The citizen science projects were crucial to preventing such ambivalent trade-offs. Their quality was largely a result of two factors: first, they were conducted over a longer period of time than the government and industry studies, and second, they included Indigenous knowledge about the "living country" they were to assess in their research design. This allowed the opposition movement to highlight two substantial flaws in the work of the proponents of the industrialization proposal: their short-termism and their lack of engagement with local actors.

Citizen scientists chose to criticize the work of the consultants on its own terms, while the *Goolarabooloo* relied on Indigenous Lore and culture when they were required to prove their custodianship in the *Bindunbur* Native Title case. They argued for the entanglement of people and the land, which they as custodians are responsible for as the reason that they should be granted the necessary rights under Native Title law. What they failed to do was to translate their interests and obligations in the land into a vocabulary of "ownership." Their opposing claimants, the *Nyul Nyul, Ngumbarl*, and mainly *Jabirr Jabirr* people, did. The latter argued for their Native Title based on the descent rule, which means that the blood lineages of claimants can be demonstrated to link up with the original inhabitants of the area prior to colonization, which in WA is the year 1829.

Similarly, citizen scientists sketched out their findings according to the expectations outlined in the law, rather than the means by which they generated them. They also applied a particular rhetoric in order to successfully contradict the alleged accuracy of the data produced by consulting firms for government and industry. Thus, on the one hand they were motivated by a commitment to "care for country" that was informed by the engagement of its Indigenous custodians with it, which is transecological and driven by lived experience, and on the other hand, they made sure that the way in which this caring for country was expressed in their reports was understandable to a Western mindset and equipped with a representationist rhetoric.

In sum, perceiving the environment as "living country" helped, both in court and on country, to evade the essentializing discourse of economic "opportunity." However, as I discuss below, while the citizen science projects were successful because they translated their results into the Western languages of law and science, refraining from doing this played an important role in the *Goolarabooloo's* failed attempt to secure Native Title.

6.1 INDIGENOUS LORE VERSUS WESTERN LAW

Twenty-five years after the first *Lurujarri Heritage Trail,* many people who since then have connected with the region in this way reunited to protest against the construction of the LNG precinct. As shown above, these included a diverse range of people: members of the wider Broome community, environmental groups, and well-known individuals from all parts of the country. Protests were further animated by casual tourists and travelers, some of them international students on a half-year trip across Australia who were eager to contribute to what they considered a meaningful cause. Others might have been "ferals," to use a colloquial term that describes a social movement mainly centered in Australia and made up of predominantly younger people who do not shy away from traveling across the country to participate in more radical forms of environmental protest and related direct action.

The participation of former *Lurujarri Heritage Trail* participants in this collaborative opposition movement provides further evidence for the agency of country. Its living heritage had attracted many to the region, even before they knew about the threat of possible industrialization. The motivations of a collaborator like Martin Pritchard when applying for his position as director of EK are a further case in point.

Though people came to defend "living country" against industrialization for diverse reasons, everyone shared an attraction to a transecology that stimulated "good feelings" in them. This is a well-known phenomenon among West Kimberley Aboriginal people, who as previously mentioned describe it using the term *liyan*. Two Australian-born student-travelers (STs) in their mid to late twenties, who had joined the opposition some months before and since then had lived with the *Goolarabooloo*, emphasized this in a personal conversation:

> *ST1:* They [the *Goolarabooloo*] really draw life from this country. It doesn't just support them in those kind of "nourishing food and water" senses. This country is their life. They get the strength to live everyday from country. And I feel that from living here. You can really feel the strength from this country around you. (Transcript 130115_TegM)

> *ST2:* The power that I felt from this country and the connection to this land, I've never felt anywhere else. The relationships that this place has fostered. . . . It's incredible and it's such a special place. But it's also coupled with this horrible environment of being in this battle constantly. It's always mixed. All your emotions are just mixed. (Transcript 130115_SLS)

With his reference to the constant battle he finds himself in, the second interviewee invokes what has been mentioned in previous chapters about how decades of mining have turned much of Australia's "living country" into "cultivated country." Walmadany/James Price Point and the coastal regions surrounding the Dampier Peninsula are exceptions to this. However, they remain under considerable threat from future development plans.

While the region's remoteness partly curtailed the exploitation of natural resources, cultivation of country was nevertheless introduced through pearling, agriculture, and livestock. The White Australia Policy was not fully applied in Broome, again for economic reasons to leave the door open for blackbirding and the continued use of an Asian workforce.[1] Before plastic was invented in the 1940s, Broome was the biggest export harbour for mother-of-pearl, and Aboriginal slaves and Asian migrants, notably Japanese pearl divers, were needed for the harvest.

The period now known as the Stolen Generations, previously mentioned by Paul N. of *Mamabulanjin*, which lasted from the turn of the century until well into the 1970s, had a further impact on the region. One in five Aboriginal children were taken from their families and sent to missions in places such as Lombadina, Beagle Bay, and Bidyadanga, where they were raised according to Christian educational and work ethics. Many who grew up in missions later moved away. Some came back when the conflict over Walmadany/James Price Point arose, and at the end of 2017, they successfully reclaimed the land that had once belonged to their respective clan groups.

On 1 May 2018, Federal Court Justice Anthony North delivered what has become known as the *Bindunbur* and *Jabirr Jabirr/Ngumbarl* Native

Title claims determination to the *Jabirr Jabirr*, *Nyul Nyul*, and *Ngumbarl* people on site, at the former Pallottine mission station at Beagle Bay (figure 6.1). The *Goolarabooloo* and others continue to challenge this claim. They argue that some have been motivated to take up custodianship not because of their *liyan* but because of an opportunity dreaming whose values they learnt to cherish as part of their Christian upbringing.

The Native Title (hereafter NT) decision remains complex and contested. Related hearings use the help of anthropologists and other expert witnesses to translate what Indigenous and non-Indigenous people present to them in so-called "lay language" into scientific reports and other forms of evidence that qualify as such. An underlying assumption of this process is that everything in and of the world can be understood, put into relationships, and ultimately decided upon by subsuming it to one narrative: that of Western law and science. Participant observations that I undertook during the final days of the *Bindunbur* NT hearings in April 2017 allow me to present this as a pertinent case in point.

One striking experience I had during those hearings was following the argument that unfolded between judge and barristers about distinctions between "rights to," "responsibilities for," and "interests in" the land. As I discuss previously, these different ways in which land—"living country"—is represented and talked about in NT points to a fundamental problem

Figure 6.1. The Altar at Beagle Bay Church. Photo: Carsten Wergin.

linked to the Settler-colonialism pillar of linear time, namely that the NT narrative wrongly confines the contemporary relevance of Indigenous knowledge to the realm of the past. This mistake has become obvious already with regard to the distinct Creole history of the town of Broome. Here, Aboriginal groups have traditionally been mixed, since well before the advancement of missions and missionaries or the atrocities committed on the Stolen Generations. Basing NT decisions on the descent rule is therefore a questionable practice and particularly so in this region.

This is a further reason why a previous NT claim for Walmadany/James Price Point had *Jabirr Jabirr* and *Goolarabooloo* working together, one providing descent and the other the continuity of Lore and culture. The WA government threatened this combined claim group with the compulsory acquisition of the land, and the initial NT claim was ended. The *Jabirr Jabirr* declared themselves in favor of the industrialization project, partly because they feared that they would lose the land anyway and so should at least receive compensation for it. Meanwhile, the *Goolarabooloo* were not intimidated in this way and categorically rejected giving up the land.

The combined claim was ended at a meeting of the *Goolarabooloo* and *Jabirr Jabirr* in February 2013 (figure 6.2). Afterward, the compulsory acquisition pursued by the Western Australian government was deemed

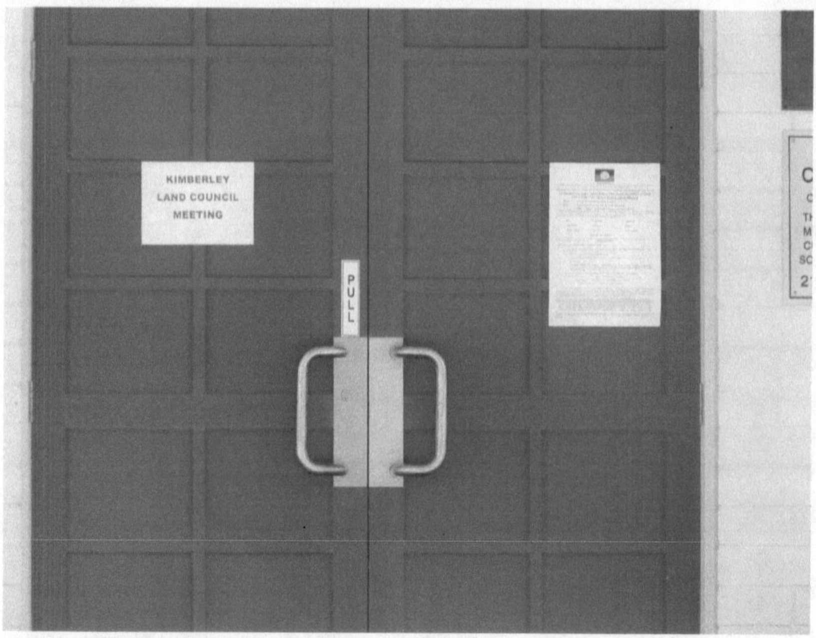

Figure 6.2. Goolarabooloo/Jabirr Jabirr Claim Group Meeting Behind Closed Doors. Photo: Carsten Wergin.

illegal by the High Court. However, a $30 million AUS Dollar compensation package remained from their attempt to acquire Walmadany/James Price Point. This package was paid out to the successful claim group(s) in the competing NT claim that followed and whose court hearings began later in 2013.

Anna worked for the *Goolarabooloo* claim and called this money "the elephant in the room" (Transcript 170404_anonymous). Everybody knew of it, but nobody seemed willing to admit how it motivated many to express their "caring for country." The ethnographic material I discuss below stems from the final four days of these court hearings on the competing claims of the *Goolarabooloo, Jabirr Jabirr, Nyul Nyul,* and *Ngumbarl.*

It was April 2017. I was in the elevator with Bart and Anna, both *Goolarabooloo,* in the Federal Court of Australia, 1 Victoria Avenue, Perth. We arrived in front of the courtroom on the seventh floor at 9.45 h, 15 min before the official beginning of the hearing. There were maybe thirty people in the audience already, plus another fifteen or so solicitors, barristers, and courtroom assistants seated with their backs turned to us. More people listened in via video link from a lecture theater at the University of Notre Dame, Broome Campus.

The barrister for the combined *Bindunbur* claim (hereafter *Bindunbur*) began his elaborations on why their claim should be given preference to the *Goolarabooloo* claim. The following five points summarize their arguments:

(1) The *Goolarabooloo* had "no rights": the *Rayi* spirit of their matriarch, Theresa Roe, does not give rights to land. Rights to the land are only given if there is proof of a blood connection.
(2) The *Goolarabooloo* had "no tradition": in the early 1930s, the late Paddy Roe [PR], and his wife, from whom the community descends biologically, arrived at a place called Willie Creek. They had five children, and therefore only the third-generation *Goolarabooloo* reaches the critical number (120–130 people) that qualifies them as a group with its own tradition.
(3) The *Goolarabooloo* had "no relation" to other groups in the area: *Bardi* and *Jabirr Jabirr* (North) are members of one society, as are *Nyigina* and *Yawuru* (South), but *Goolarabooloo* are not such a local group.
(4) The *Goolarabooloo* had "no succession": even if it can be argued that the *Jabirr Jabirr* have foreshadowed their extinction in order to have their country looked after, if their children come back from wherever they have been, there cannot be (even partial) succession.[2]
(5) The *Goolarabooloo* had "no evidence": all depends on the factual truth of PR's narrative about the actual extent of extinction.

The *Bindunbur*, unlike the *Goolarabooloo*, are blood-related to the people who inhabited the land in question prior to colonialization. *Goolarabooloo* Lore and customs are furthermore closely tied to those of the *Nyigina* people, since PR and his wife descended from them. However, they only came after the *Jabirr Jabirr/Bindunbur*.

The question of whether this transition from *Jabirr Jabirr* Lore and customs to a hybrid version of those practiced by the *Goolarabooloo* legitimizes their succession as rightful claimants to the land was translated into an argument of rights to the land versus responsibilities for it. Rights attributed due to a proven blood lineage that dates back to the time before colonialization are one significant component in justifying Native Title according to Western law. However, viewed through the transecological lens, this introduces linear time as one of the central components of the decision-making process—one of the four pillars of Settler-colonialism. For obvious reasons, the *Bindunbur* adhered to this law, while the *Goolarabooloo* emphasized ongoing intra-action in "living country." In other words, the *Goolarabooloo* presented a transecological view through which their actions on country provided sufficient evidence for their custodianship. In contrast, the *Bindunbur* claimed to represent what in Povinelli's terms are "the real natives."

Emphasis on the here and now, however, is not accommodated in NT law. NT law is set up in order to give back to its rightful owners land that was wrongly taken away based on the (deliberately) false assumption that Australia was a *Terra Nullius*, a land that belonged to no one. NT legislation therefore seeks to reinstate a situation that existed in the past, to the extent that the present allows for this. However, this leaves insufficient room to accommodate the transecological case made by the *Goolarabooloo*, which strongly relies on the acknowledgment of what is in place—the current situation and the lived experience of it. In other words, while the *Bindunbur* have rights to the land according to NT legislation, in light of their entanglement with it, the *Goolarabooloo* have the custodial responsibility. As will be seen below, however, the latter cannot be accommodated by NT law.

Both groups are for different reasons legitimate claimants to the land. Due to the impact of Settler-colonial legislation and the related killing and/or forcible removal of Aboriginal people, in particular children, it has not been possible for many to continue their intra-action with "living country." Meanwhile, many also seek to recommence this intra-action, to the extent that it is still possible. In this regard, the *Jabirr Jabirr* had an advantage in that in their absence the *Goolarabooloo* had taken up the responsibility of caring for "living country," of intra-acting, of continuing cultural practices, and of preserving the living heritage of the Northern Tradition for which they are *Maja* (Law Bosses). On another front,

Goolarabooloo actively encouraged others, whether non-Indigenous people or *Jabirr Jabirr*, to join them.

This collaborative scenario provided for the combined claim of *Goolarabooloo* and *Jabirr Jabirr*, but that claim ended because of their dispute over the industrialization of Walmadany/James Price Point. In light of the outcome of the *Bindunbur* case, although construction of the LNG facility was abandoned, the project proposal has therefore been indirectly accountable for a partial cultural genocide. The *Goolarabooloo* have literally lost the ground on which to perform their cultural practices, and at the same time it has become an impossible task for the *Jabirr Jabirr* to reinstate their Lore and culture without the *Goolarabooloo*. Both groups are weakened because the NT decision has separated them, which in turn has opened the door to future industrial development on the land.

To this day Theresa Roe identifies as both *Jabirr Jabirr* and *Goolarabooloo*. Nevertheless, the *Bindunbur* argued that her connection to the land through a *Rayi*, a children's spirit that came to her parents via a stingray does not give her rights to it. From their point of view, her *Rayi* marks a personal relation to the land, not a right to it, because a right needs to be possessed under the "normative system." Following on from this, the *Bindunbur* argued that initiated men share responsibility for the Song Cycle that traverses the country, but that no rights are connected to that responsibility.

In sum, the *Bindunbur* recognized individual relations between people and country because of their *Rayi* and/or customs, but not the *Goolarabooloo*'s right to the land as a group, because such rights must be related to reciprocal relations with other groups in the area. Here, the *Bindunbur* followed the anthropologist Peter Sutton, an expert witness heard in the case, who argues for a people–country-language nexus. In this argument, neighboring groups (i.e., *Yawuru, Ngumbarl, Jabirr Jabirr*) have a right to take over country, which is commonplace across Indigenous Australia where NT determinations are regularly made according to language groups.

Back to the courtroom. At this stage in the hearing, the *Bindunbur* had reached what is likely to be the most sensitive issue, namely the "truthfulness" of the narratives of Paddy Roe (PR). It is worth returning to the notes I took from the dialogue between Justice North and the barrister for the *Bindunbur* that unfolded on the issue:[3]

> *Justice North:* PR never claimed for himself that the land was given to him as the owner. PR only ever made the claim to be a caretaker. So you cannot criticise him for claiming something like ownership.
>
> *Bindunbur:* There is an earlier version of the PR story that differs significantly, which leads to the conclusion that even the idea that he was given the land to take care of is questionable.

JN: That would require the conclusion that he was making up the story. A very harsh finding to make.

B: PR was not consistent enough to present a balanced story in his accounts. And there is pattern enough that the communication was directed all in one way, possibly to achieve a particular result: Stories were most often told to white men, government officials, rather than to the people who his stories acknowledge as their country. So the PR story can certainly not be accepted as the basis for rights and ownership in a country (. . .)

JN: When is the first external sign of PR taking a caretaking role?

B: It might be Prof Muecke's work in the early 1980s. There is also evidence of Mrs Roe [Theresa Roe].

JN: So it is about 40 years after the event that any publicity emerges apart from looking after the old people (. . .) Prof Muecke's involvement arose from his own academic interest?

B: He clearly regarded himself as useful to PR to share the story with a white audience.

JN: That was when the land rights era emerged.

B: The question is what was PR's response to this: was he still simply taking care or was he starting to position himself in relation to the land rights debate in Broome?

JN: So what was his agenda? What do you draw from this? First, "I like this country." Then what, "I'm positioned to make a claim given the new legal environments?"

B: He clearly told his children and grandchildren these stories.

JN: If you are asking me to find that the story was invented, then this would only be in conjunction of finding that it is for a purpose (. . .)

B: The Roe family stands against evidence on the *Bindunbur* side to never have heard of the story at all.

JN: There is potential of doing great injustice to PR and his family if I make a finding that advocates such a thing (. . .) Do you want to fight an intertribal battle in the courtroom? Or do you want to provide proper evidence? You need to pin your case very clearly. Was PR a charlatan and a liar? (. . .) We have evidence from the grave [refers to audio tapes of interviews with PR] of a central issue, "Here is what he said." Do you really want to get from this to disprove that? (PAUSE)

[Bindunbur asks for time to reflect on this.]

JN: Whatever PR said, whether it's made up or not, does not play any role under the NT act. The question is: what is the source of that Lore?

Paddy Roe was a man of great standing in the Broome community at a time when the socio-political environment was hostile toward Aboriginal people, enforced for example by a prohibition on being inside the Common Gate after sunset, the forcible removal of children from their families, and blackbirding. Roe's vision was of protecting Indigenous culture and living heritage by educating people about it—relevant decision-makers in particular. His advantage was that he was a charismatic and sought-after person for many: government officials, media representatives, and academics as well as casual tourists and travelers. If ever one asked in town about who to contact in order to learn more about Aboriginal issues, people would point to Paddy Roe.

Meanwhile, the exchange between the *Bindunbur* barrister and the judge about the truthfulness of PR's story was the closest the court came to any recognition of "living country" in transecological terms. The *Bindunbur* suggested to the court that it make a finding that would label PR a charlatan and a liar. They sought to stabilize their claim using another of the four pillars of Settler-colonialism, which is naming—in this case equivalent to shaming. However, the judge rejected this because of his awareness of what Paddy Roe and his family have done for the Aboriginal communities in and around Broome. For this, Justice North relied on sound "evidence from the grave," as he termed it, which was the closest he could get to the lived experience of PR as informant, custodian, and *Maja*.

Nevertheless, at the end of his exchange with the barrister, Justice North stated that he cannot base his decision on the question of truthfulness, but on the question of whose Lore prevailed on the land prior to colonization, that is, before 1829. Thus it was clear by this stage that the *Jabirr Jabirr* are to be given Native Title. The only question remaining was what future role, if any, the *Goolarabooloo* would be given in this scenario.

In response to the exchange between Justice North and the *Bindunbur* barrister, on the following day the *Goolarabooloo* barrister stated that Richard Hunter, born in 1957, recalled his grandfather telling him stories about *Goolarabooloo* country when he was a child. This was at a time well before NT was even imaginable, let alone rights for Aboriginal people to reclaim their land. The reasons for telling these stories are related to the preservation of "living country" and its living heritage through stories passed on from generation to generation, as Aboriginal people have done "since day one," and are not motivated by new legal environments. As the barrister made this point, Richard Hunter, who sat near me in the audience, nodded for a long time.

The *Bindunbur* refrained from pursuing the question of truthfulness further but insisted that (1) the *Goolarabooloo* were "new" people to what for them was a new area, and (2) their Lore status, a hybridized version of *Nyigina* and *Jabirr Jabirr* Lore and Theresa Roe's *Rayi* were not enough for them to be the subject of determination of NT. On the contrary, *Goolarabooloo* Lore and customs had no roots in the area of the claim. This meant that if *Jabirr Jabirr* and *Goolarabooloo* were given the claim, two sets of Lores and customs would operate simultaneously, which had not been the case prior to 1829.

The above presents a rough summary based on notes I took of what was presented by *Bindunbur* over the course of nearly a full day. They are my interpretations of what I witnessed in court and any misrepresentations are fully attributable to me.

The following day, the *Goolarabooloo* barrister (hereafter *Goolarabooloo*) continued with his presentation, which included what for me became another central dialogue between him and Justice North about the notions of "rights" and "ownership":

> *Justice North:* Theresa's *Rayi* gave her right to ownership, that is what you say?
>
> *Goolarabooloo:* We would not claim ownership.
>
> *JN:* But you say that the rights of her through *Rayi* give her descendants the rights to the area.
>
> *G:* We say Law Men are given rights to protect.
>
> *JN:* But not rights of exclusive possession?
>
> *G:* No.
>
> *JN:* I misinterpreted the case then.

The *Goolarabooloo* unfolded their argument around the terms "rights," "responsibilities," "interests," and "ownership." They admitted the point that rights that are given through *Rayi* and custom are different from rights that are given through story or Lore-related objects. The crucial difference was that *Goolarabooloo* pronounced the claim area to be a transitional zone on the Song Cycle, between the Northern and the Southern Traditions. They considered the whole of the Dampier Peninsula as one society, with transitional areas and changing environmental conditions that included all groups. Accordingly, they argued that the transitional area that the *Goolarabooloo* had been looking after should not be allocated to one group based solely on descent. In addition to this, they argued that it was commonplace for such overlapping areas to accommodate more than one Lore.

The definition of the Dampier Peninsula as one "society" meets the criteria of other expert witnesses, in particular those of anthropologist Geoffrey Bagshaw, for whom a society is not a bounded or sealed unit but one whose Lores and customs, such as marriage patterns, trade, mythologies, and ritual relations, are shared by a wider group of people. Sutton also sees this society as extending beyond the peninsula to include the *Nyigina* people because of a shared belief in *Bugarrigarra* and a shared language.

Furthermore, the *Goolarabooloo* pointed to the fact of there having been a single system of religious Lore in place on the Dampier Peninsula in 1901, when data were collected by Daisy Bates on the "Koolarrabulloo [*sic*] tribes of Broome" as evidence for such ritual syncretism (Bates [1947] 1985, 22). There is therefore agreement among anthropologists on the existence of a transitional region somewhere in the area. According to anthropologist Scott Cane, who bases his findings on Bates, this encompasses *Nyigina* Lore and custom, which proves that the south-western part of the claim area, which is the area that the *Goolarabooloo* claim focused on, includes *Nyigina* people. "And," the barrister added, "of course PR is a *Nyigina* man."

In other words, while both barristers agreed that succession is more likely to be accepted between groups that are not strangers, the *Goolarabooloo* argued, contrary to the *Bindunbur*, that the *Nyigina* were not strangers. In addition, they pointed out that not every group or area adheres to all the Lores and customs of the Dampier Peninsula society. Thus, interest in a particular area might be different from a right to it, but it still needs a normative/customary basis because there is more than one interest and more than one group "interested" in this particular area, which is transitional.

In sum, the *Goolarabooloo* termed their claim an "area of blending." Their current generation are descendants of Paddy Roe, albeit with what Cane terms a "shallow generational memory" that is particular to the Dampier Peninsula. Furthermore, Paddy Roe and his story of being given the land to care for had been accepted until the Walmadany/James Price Point conflict elevated it to a point of dispute. Before this, the *Goolarabooloo* and *Jabirr Jabirr* shared a combined claim, and Paddy Roe would have succeeded in it as a descendant under the descent rule.

Another point made by the *Goolarabooloo* was that the position of PR within Dampier Peninsula society had been confirmed by further evidence given about Lore meetings he participated in, where seating arrangements show social and societal acceptance. These meetings restated the point that rights in particular areas might not be shared, but that what is shared is a risk of being harmed if significant objects and Lore grounds are damaged. This is the reason why the *Goolarabooloo* felt the absolute need to defend Walmadany/James Price Point from industrial development—because of their obligation to care for its living heritage.

The *Goolarabooloo* argued that (1) the claim area is a transitional zone and as such should not be allocated to one group solely based on their descent, and furthermore that (2) ceremony is a collaborative exercise on the Dampier Peninsula, while the protection of Lore-matters is not. Therefore, "obligation" and "interests" were better terms than "ownership" to define the relationship of the *Goolarabooloo* to their claim area.

The above argument highlights ontologically different world(view)s existing parallel to each other, the one of *Bugarrigarra* Lore and the one of Western law. As became evident when Justice North refused to engage in the discussion about Paddy Roe's truthfulness, the "NT narrative" allows for an imaginary that caters for more than one of these worlds. However, Settler-colonialist understandings prevail when it comes to making diverse voices heard. The reason for this is that NT demands that people attest to a time prior to colonization in order to make a legitimate claim. Such a demand confines Aboriginal people and their Lore to the realm of the past. As Stephen Muecke argues, "NT law tends to sever the fundamentally important Indigenous link between body and country to (as Bronwyn Lay brilliantly puts it) 'freeze the continuity, reciprocity and growth of Aboriginal jurisprudence' by putting it in the past as tradition, or 'normative system'" (Muecke 2017, 140).

In NT, recognition is granted if one is able to show that one is a "real" native, somewhat frozen in time. Some Indigenous groups find economic opportunities in the fulfillment of related imaginaries. The point made by the *Goolarabooloo* about their claim area being a transitional zone challenges this imaginary and highlights the pitfalls of NT law in as much as it turns the controversy between *Bindunbur* and *Goolarabooloo* over the recognition of their NT claims into a controversy over the recognition of ontological differences. The question became: Should the transecology of "living country," the entanglement of people and place, and its living heritage that is continuously reinstated through lived experience be defined through abstract representationism, or real-world experience?

The *Goolarabooloo* dared to try for the lived experience and for what is in place, because that was what they had learnt and practiced for over thirty years together with participants in the *Lurujarri Heritage Trail*. The *Bindunbur* opted for representationism and what was in place in the past. The latter won, because anything else would have challenged the very foundations of NT law. The *Bindunbur* developed their argument according to the Western law model and were successful, while the *Goolarabooloo* remained earthbound and lost. The recognition of their custodial status might well have been a door-opener for the Australian Dreaming I draw attention to here. Instead, it produced another failed attempt to reconcile Settler-colonialist Australia with the Lore of its First Nations.

6.2 SCIENCE CONSULTANTS VERSUS CITIZEN SCIENTISTS

The above accounts for the continuous drama caused by colonialism and also gives proof that the project of decolonization is far from accomplished. The devastating consequences of colonization are felt on a global scale in the Anthropocene/Capitalocene era. What is needed is a new narrative, which I argue is readily available in the Australian Dreaming described here. Citizen science projects conducted as part of the opposition movement against the industrialization of Walmadany/James Price Point add a further dimension to the powers of this Australian Dreaming, which draws on the earthbound experience of "living country."

Previous scientific studies have recognized the Dampier Peninsula as a migratory pathway and habitat for a great variety of endangered animals such as whales (Jenner, Jenner, and McCabe 2001), turtles (Mustoe and Edmunds 2008), and dugongs (Marsh et al. 2002). Based on these and other studies, the coastline of Walmadany/James Price Point has been termed a "high marine productivity location" that gives rise to an abundance of fish, which in turn supports an abundance of seabirds and other predators.

A joint report by environmental organizations The Wilderness Society, Save the Kimberley, and Environs Kimberley identified more baitfish off Walmadany "than on the Great Barrier Reef" (The Wilderness Society et al. 2012; see also Cappo et al. 2011). Citizen scientists for the preservation of Walmadany/James Price Point found a nest of the critically endangered hawksbill turtle, evidence of the first possible hawksbill hybrid in WA, and more turtle nesting sites than had been reported in the WA government's 2011 "Browse LNG Precinct Strategic Assessment Report" (hereafter SAR) (WA Government 2011).

Despite their efforts and those previous findings, on 16 July 2012, the WA Environmental Protection Authority (hereafter EPA) followed the SAR in its decision to recommend the LNG facility to be built. Next, I present ethnographic material that I recorded during an event that took place on 26 July 2012, ten days after this recommendation was made public. It stems from a meeting organized by Environs Kimberley and the No Gas Campaign and held at Broome Lotteries House to inform the wider community about the findings of their citizen science projects and on how to formulate appeals to the EPA's decision.

The goal was to inform the general public and motivate as many as possible to send appeals to counter the decision. One piece of advice that citizen scientists gave was for community members to state in their appeals that "the inability of the SAR researchers to find bilbies indicates that their terrestrial fauna studies were poorly replicated temporally and spatially" (Transcript 260712_anonymous). In the SAR, the absence of evidence for bilbies was assumed to be evidence of the absence of that species. However, if findings of citizen scientists had been taken into account, this absence would have been dismissed as the result of an imprecise research model.

The EPA, however, preferred SAR's "alternative facts" to the results of the citizen science projects. Frustration and anger arose because of this, but the Lotteries House event was still held in a positive spirit. Like the concerts of the Pigram Brothers on Male Oval and Cable Beach, or the Broome families event in the churchyard, Lotteries House was a gathering held not to mourn the negative decision announced by the EPA but to celebrate the quality of the scientific studies that the wider community had been able to produce. As I demonstrate below, this quality was largely due to the fact that the research undertaken was "earthbound" in more than one way. Citizen scientists, as much as their results, not only stemmed from and were informed by "living country" but also stayed in the community.[4]

The presentations that evening at Lotteries House discussed findings on turtles, dolphins, whales, and bilbies. The latter are members of a group of ground-dwelling marsupials known as bandicoots; they have long pointed snouts and compact bodies, measuring between 29 and 55 centimeters in length and differing from other bandicoots in having longer tails, long silky fur, and large ears. Because of their looks, Australians have adopted them as a legitimate alternative to the Easter Bunny, not least because rabbits have caused significant damage to the Australian environment.

Bilbies used to inhabit more than 70 percent of Australia, but the population of the only remaining member of this species, the Greater Bilby, has been reduced to inhabiting less than 20 percent. Greater Bilbies are mainly found in areas that have undergone fewer industrial and agricultural transformations. The 2011 SAR concluded that there was virtually no evidence of bilbies on the Dampier Peninsula. However, Indigenous custodians had repeatedly confirmed sightings and encouraged a search for them. The result was the report "Evidence of the Greater Bilby (*Macrotis lagotis*) at the site of the proposed James Price Point Browse LNG Precinct" (Goolarabooloo et al. 2011), published more than nine months before the EPA's decision in favor of the LNG facility.

The report was the product of long observation periods conducted in and around the area of Walmadany/James Price Point. What is of particular relevance to the argument put forward here is people's motivations for participating in these surveys. These were fundamentally attached to the experience of "living country" and the Australian Dreaming it pertains to. Below, Chris D., a marine biologist, describes how he arrived in Broome, connected with the *Goolarabooloo* through, you guessed it, the *Lurujarri Heritage Trail* and eventually became a substantial collaborator in the preparation of this report:

> *Chris D.:* The reason I'm in Broome is because of my partner Bernadette who did [the *Lurujarri Heritage Trail*] as an RMIT student 12 years ago and really loved it, met the family and then started coming back (. . .) I was very keen to

see that part of Australia, so I nagged her to bring me up here (. . .) That was five years ago [2008], and that was just hanging out at Coconut Wells [*Millibinyarri*] . . . just camping there at the bottom of the Trail camp. We went up to Yellow river [*Bindiangun*] with [Mr Roe] and Margie [Mr Roe's wife] and the family (. . .) Really amazing, mind-opening first encounter for me. . . And that generosity that they showed . . . I was just Bernie's partner and all of a sudden they are taking me fishing and everything... Was really nice to see. And then we came back the next year (. . .) and that's when the gas plant was really starting (. . .) I was doing a PhD at the time so [Mr Roe] was asking, "So you're a marine biologist are you? Do you have your ticket yet?" "My ticket . . . Well, I guess I'm still in my PhD . . . so I guess I don't technically have my ticket." So he said, "Once you finish your ticket come over here and help me because I want somebody to be able to read the scientist reports that Woodside and the state government are doing and tell me if they are truths . . . accurate and all that stuff."

And, then the next year we came up (. . .) I think it was July 2010 (. . .) I came up and waved the idea to volunteer on Trail (. . .) We were only going to spend one or two months up here but because of what happened with Black Tuesday we were getting involved in that campaign a lot more and ended up staying for five-six months, doing multiple trips back to Melbourne and Broome (. . .) In early 2011 we organized a big fundraising concert where we flew down [Mr Roe] and Steve Pigram and had a lot of Melbourne bands, which was really good (. . .) Bernie was putting up work and putting up other projects she had. I was taking time with my PhD. We were spending our savings and all that stuff to be up here.

Carsten Wergin: What attracted you?

CD: It was really the cultural aspects (. . .) In terms of the actual environmental aspects, although they are fascinating and really incredible, I've got a far greater knowledge and understanding of the bush and marine environment around Victoria and there's plenty of incredible places down there which are also in great need of protection. So in terms of helping out the campaign (. . .) it was actually the cultural aspect, the *Goolarabooloo* and their integration with the environmental stuff, but also how willing they are to share their culture through the Trail (. . .) They knew the boundaries so therefore they were able to really interact a lot more with non-Indigenous people (. . .) And that initial kind of introduction, through the *Goolarabooloo*, through the colleagues going out to other Indigenous communities and other ecological areas around the Kimberley, has really set me up quite well. And that's one of the reasons why I feel like I have been mildly successful in my job, where we're working with Indigenous rangers, because I have that schooling. I'm not just a fresh *kartiya* [whitefella] from the south, who's been to uni, "He's got all the qualifications and doesn't have much interaction with Indigenous communities . . ." The fact I actually do have schooling and learning from Richard and from Frans or Phil, from [Mr Roe] and the family . . . be it just the daily interactions and how value systems are so separated. . . . Even coming down to the ecological values. . . . The only reason I know anything about

> monsoonal vines thickets is because of what Richard and Frans have taught me. (Transcript 130513_1_ML)

Chris D. names the cultural aspects he encountered with the *Goolarabooloo* as the thing that motivated him to contribute to the opposition movement. His introduction to these began with his participation in the *Lurujarri Heritage Trail* and the sharing of country with the *Goolarabooloo* families. This is what "set him up quite well" and what allows him to "feel like I have been mildly successful in my job."

The significance of his assessment lies in the fact that it was not his scientific qualification that provided him with this feeling, but the shared experience of "living country" and exposure to Indigenous world(view)s—to value systems that are, as he states, "so separated"—which eventually motivated him to move his center of life to the Kimberley. His commitment to "living country" became existential to the point where he and his partner were "putting up" the life they had in Melbourne, delaying or even partly abandoning the life course they had envisioned for themselves and spending all their savings to be there for it.

At the time of writing, Chris D. is still in Broome. He "got his ticket" and secured a position with EK as coordinator of their Kimberley Nature Project. He contributed to the Lotteries House meeting in Broome on 26 July 2012 as one of the collaborators in the citizen science projects, who explained to the audience that observations of bilbies "on country" began with the sighting of "strange tracks that looked like bilby tracks" (Transcript 260712_anonymous). Based on those sightings, citizen scientists observed the dirt road on which the tracks occurred. They camped "on country" and swept the road each evening. Each morning, they found new tracks, and shortly after, discovered a bilby burrow. In a later interview, he explained this collaborative work in more detail, and how it sharpened his views on ecological research and the transecological dimensions of it:

> Damian and various other people on country were trying to do fauna surveys (. . .) I started talking to Damian and Shawny Bob (. . .) Louise [Middleton] had done a weekend survey with the Skippers, which is a community plant and animal group up here. So they had done a bilby survey (. . .) Phil had seen a bilby near Quondong, I think in 2006/2007 (. . .) So a lot of the *Goolarabooloo* thought they'd be out there. . . . But also seeing a lot of the work that goes down for ecological consultants for Woodside and the government and seeing how, at times, inadequate some of those traps [were]. . . . A lot of those people thought that if there's bilbies out there they wouldn't have found them (. . .) They weren't staying out in country (. . .)
>
> Ecology, unlike other science, is very observationally based, is less theoretical than other sciences. So, hunches and feelings and getting to know an area is actually quite important for ecology. It's not something just spoken about

> that much. If you stay there and you camp there and you start to feel for the country, all these concepts which Indigenous people talk about but scientists don't really talk about. There's this aspect of the *Goolarabooloo*, of their traditional ecological knowledge, which just absolutely takes your breath away (. . .) There's always incredible information which would help non-Indigenous academic ecologists (. . .) Why is there a difference? Why aren't we as good at reading the country?

The combination of Indigenous means of "reading the country" with Western scientific expertise and technologies made the difference when it came to the quality of the citizen science projects. Motion- and heat-sensitive cameras were installed next to the bilby burrows located by the citizen science team in hopes of capturing a picture.[5] Meanwhile, they continued their observations south of the dirt road, where they found another forty recently used bilby burrows, which meant an active bilby area. It took six weeks for a picture of a large adult in good condition to be taken. More photos and video footage followed, garnering images of three adults and their young jumping around in front of the camera.[6] As bilbies have a maximum of three young, the images confirmed that Walmadany/James Price Point and its surroundings are prime habitat for these animals.

Figure 6.3. No Gas Campaign Meeting at Lotteries House. Photo: Carsten Wergin.

The main reason given by the citizen scientists that evening at Lotteries House as to why they could provide such different results from the WA government surveys was the significant difference in study length. Citizen scientists had lived and worked on "living country" for two consecutive months. In contrast, the consultants for WA government and industry took a maximum of two and a half weeks to complete their studies. More time spent not only on but with "living country" led to more accurate results.

A further difference lay in study focus. While the citizen scientists concentrated on bilbies alone, the consultants also studied other mammals—all fauna in the area, even. Such all-encompassing aims diluted their findings and negatively impacted their efficiency and accuracy. This is linked to the fact that their sources of information about where to make possible findings were limited to government data about previous studies in other parts of WA. While the consultants relied on those to determine possible habitats, citizen scientists drew on Indigenous knowledge and their means of reading the country. Chris D. explains this further with regard to another citizen science project conducted on turtle nesting in the vicinity of Walmadany/James Price Point:

> They [consultant firms] weren't just going to survey the area right next to the gas plant. They were going to survey buffer zones around it, so 20 or 30 kms coastline in the end (. . .) There was no input from the Traditional Owners [TOs], I don't think, so in the end the scientists just made the call on what they thought was appropriate habitat (. . .) And then they started (. . .) saying, "[Walmadany / James Price Point] is really kind of rocky, pebbly beach, the turtles aren't going to nest there. If there's scraggly reef in the shallow water the turtles (. . .) won't like that so they will swim around it." They started determining what they thought was appropriate turtle nesting habitat, which is a very normal process to go through as a scientist. You're limiting your survey area. In the end where they looked was not where the turtles nest, so they didn't find any turtles nesting. For *Goolarabooloo* that was almost laughable because they know there's turtles there. They—in the past more so but occasionally still do—eat turtle eggs from the nests.

> Once again, through the *Goolarabooloo*, the main people who really got that up and running as initial thing were Ty, and Jules and Nadia, as well. They were the ones going, "We're going to do an internal survey." And at first I was a little bit, "Sounds like a lot of work and not enough good results."Anyway, we then had a workshop... And Jules is very important, Jules and Kevin Smith, because they've done turtle surveys for Conservation Volunteer Australia. . . . So we all go together. We had Phil and Richard there and they said to us, "Look, the only area you need to look at is between James Price Point and the black sands gully to the North." (Transcript 130513_1_ML)

Again, incorporating Indigenous knowledge about the area made the difference, as did a collaboration between knowledge holders and "passionate volunteers," quite a few of whom had participated in the *Lurujarri Heritage Trail*, experienced "living country" first hand and become part of its transecology. They were people who had learnt to enjoy being "on country," which provided energy and *liyan*. As a result, they not only worked *pro bono* but did so over a period of months rather than a few weeks.

If a similar effort had informed the SAR, the costs for the WA government would have been significant, if only in terms of the additional hotel bookings, car hire, and salaries for scientists and consultant firms. What would have been even more challenging than such financial constraints, however, is that the consultants would have needed to connect with country in a similar way to the citizen scientists. They would not only have been required to spend more time "on country," to acknowledge Indigenous knowledge about "living country" in their research design, but also to seek their own entanglement with it. They would have needed to strip bare, leave their conventional approaches behind and accept their work as part of a wider transecological setting.

My personal account of the challenges I faced when writing a report about sacred sites in the dune system, which is given in the following chapter, shows how Indigenous knowledge is not representable in Western scientific terminology alone. Such "white magic" is only magical to the extent that it renders significant aspects of reality invisible. The same accounts for the writing of scientific reports on bilbies and turtles. The Western approach taken by the consultancy firms rendered significant aspects of "living country" invisible, as it did not require them to intra-act with its living heritage. This is why the surveys for the SAR could conclude that the bilby population in and around Walmadany/James Price Point was not significant. Again, Chris D. on this point says:

> The scientists (. . .) who were working for Woodside (. . .) They took their TOs with them (. . .) They were typically younger traditional owners, younger guys who hadn't spent much time out there (. . .) There was a bilby survey going around the trace line, which is the road, which was going to be proposed to go to the back of the gas plant. Phil asked me to go out there to check and make sure that these ecologists working for Woodside weren't doing anything dodgy (. . .) We found them and I spoke to the ecologist (. . .), "I'm not going to get in your way (. . .) Do you mind if we tag along?" And they were fine with that. They had two TOs with them and a Woodside employee. At one point one of the guys got very angry because we were there. . . . Started swearing, getting really aggro (. . .) In the end he calmed down and he said that for him it was an expression of anger at the issues and (. . .) the devastation that had happened to the community because of

> the gas plant. His family were in favour of the gas. They grew up with the *Goolarabooloo*. They were mates from the north (. . .) I started having a chat with him, "How long you been working for Woodside?" "Oh I don't know, one or two years." "How the work been?" "Been really good. Last year we went out with anthropologists through the dunes, it was really good learning from them about how the old people used to live."
>
> For me that just really . . . I mean, there's a lot of Indigenous communities where they have to learn through anthropologists or anthropological reports and that's unfortunate. . . . But being in this situation where you have people like Frans and Richard and Phil and [Mr Roe] who know everything and have only led out a small percentage of what they know . . . learn through anthropologists . . . [interrupts sentence shaking his head]. (Transcript 130513_1_ML)

It was Western science that provided the sole basis for the EPA's recommendation to develop the site (The Wilderness Society et al. 2012), all the way down to the fact that information that had been provided by Western scientists (anthropologists) to TOs was then filtered back through to the scientists to produce the environmental reports for Woodside. Meanwhile, everyone present that evening at Lotteries House acknowledged the agency of "living country" and the values of its living heritage. Various participants in the citizen science initiatives termed each other "lucky to actually find the bilby colony when and where they did." The *Goolarabooloo* took this one step further and affirmed that "country put them there." Their collaborative knowledge made the difference, as Chris D. points out with regard to his own role in the surveys:

> What I brought to the team out there was doing the *kartiya whitefella* stuff. Marking stuff on the GPS, taking notes, drawing maps, all that stuff, and in the end writing up the report and handling all the media and interviews and talking with cameras about what that means (. . .) We needed both! If it was just Cory and Damien then maybe they would have found bilbies. [But] they wouldn't have known exactly how to write up a scientific report that (. . .) the people can take seriously. They wouldn't have known how to talk to the journals, "Plant Cell," "Ulysses" or so on (. . .) Me by myself, although I might have found bilbies, having their skills definitely increased how many bilbies we've found. . . . And the knowledge we've built on them, too. . . . And then put into the scientific reports. (Transcript 130513_1_ML)

Goolarabooloo could also have easily indicated to the government and industry consultants where bilbies or turtles are to be found. However, such collaboration would have jeopardized the development plans in more than one way: (1) it would have made the research significantly more cost-intensive and (2) it would have required a recognition of "living country" and of Walmadany/James Price Point in transecological terms.[7]

Instead, in the assessment that informed the recommendation of July 2012 in favor of the LNG precinct being built, the EPA turned the citizen scientists' argument around. It concluded that there must be even more bilby habitat on the Dampier Peninsula than what had been found.[8] This is in line with general attempts to downplay ecological destruction in a mismatched calculation in favor of possible economic gain. I shared a related experience during the 2012 *Lurujarri Heritage Trail*, when one of the guides encouraged us to fish near the rocks at Walmadany/James Price Point. The place is a preferred fishing spot. However, Woodside had installed a rig offshore to conduct exploratory drills for the construction of a two-kilometer-long jetty to accommodate LNG vessels. The sound these drilling activities made was barely audible above sea level but had impacted under water to the extent that we were unable to catch anything that day.

This brings me to the third and final example of a successful citizen science project, which was conducted on arguably the most impressive mammals to migrate along the Walmadany/James Price Point coast: humpback whales. With between 26,000 and 34,000 individuals per year, the humpback whale migration from Antarctica to the summer feeding grounds along the western coast of Australia accounts for the largest population of migrating whales in the world, and the animals are

Figure 6.4. Drill Rig at Walmadany/James Price Point. Photo: Carsten Wergin.

listed as vulnerable by the Australian government. In the course of the citizen science initiatives, the Kimberley Community Whale Research Project formed and conducted a three-month, land-based humpback whale survey. It investigated humpback whale numbers, behaviors, and habitat use of the area in order to assess the impact of the proposed LNG development.

Meanwhile, Woodside had commissioned a company called RPS Consultants to survey the humpback whales, a survey on which the EPA later based its approval. That evening at Lotteries House, citizen scientists explained that the consultants had estimated that during the 2012 migration season approximately 5 per cent of the total population, or 1,000 whales, would pass within 8 kilometers of Walmadany/James Price Point (Woodside 2012). The eight-kilometer margin determined the number of whales who within this boundary would be impacted by the heavy ship movements that would accompany the LNG processing hub. However, not only did the consultants' survey account for a remarkably low number of whales within this boundary, it also suggested that the site for the LNG precinct was generally less attractive as a resting place for humpback whales, since it was located on a straight section of coastline (figure 6.6).

Community members expressed their concern about the scientific rigor of these findings and joined forces with national and international volunteers to produce their own study. As with the research on bilbies and turtles, participants in this citizen science project again contributed their time voluntarily. They came with varying degrees of expertise, from people interested in conservation to work-and-travel visa holders, science students, and marine biologists.

Again, study length made the difference. RPS Consultants conducted ten aerial surveys throughout the migration period. Citizen scientists lived and worked on "living country," watched every day for four hours, from 8 am to 12 pm, over a period of three months, with six volunteers on each of their two platforms on a cliff top at Murdudun, eight kilometers south of Walmadany/James Price Point (figure 6.5). Their survey area covered 223 square kilometers of the region. A pearl-fishing boat sat offshore and was used to measure how far whales were from the coast.

I was a participant in the 2012 *Lurujarri Heritage Trail* when I took the photo above (fig. 6.5). Melissa Parke from the Australian Labor Party is seen (second from left), with her back turned to the camera and in conversation with one of the citizen scientists. She had come out for the day in the course of another visit to Broome. It was a mind-opening experience for me to be at once part of the Trail experience and at the same time witness to how Australian politics was catered for in order to assist in the recognition of the particular values of "living country" that were under jeopardy.

At the Lotteries House meeting, the group was three weeks into their whale research, and a citizen scientist outlined preliminary findings. They had followed the question of how many whales would travel within eight kilometers of the coast, and also what their behavior was. Were they traveling through, or were they feeding, playing, even calving in the area? During these first three weeks, citizen scientists spotted 767 whales less than eight kilometers from the coast, including newborn calves. Since whales travel day and night, this suggested that during that three-week period, a total of approximately 4,500 whales had already passed through Walmadany/James Price Point. Their results were in stark contrast to the numbers estimated by RPS Consultants, who predicted 1,000 whales for the whole migration period.

In addition, citizen scientists found that humpback whales can calve anywhere along the coast north of Broome, including significantly closer to Walmadany/James Price Point than was presented in Woodside's results.[9] The latter argued that 95 percent of the whales moved outside the eight-kilometer range, while the citizen scientists found many of them moved between four-and-a-half and eight kilometers from the coast, depending on whether the tide was high or low.

Figure 6.5. Whale Watching Platform at Murdudun. Photo: Carsten Wergin.

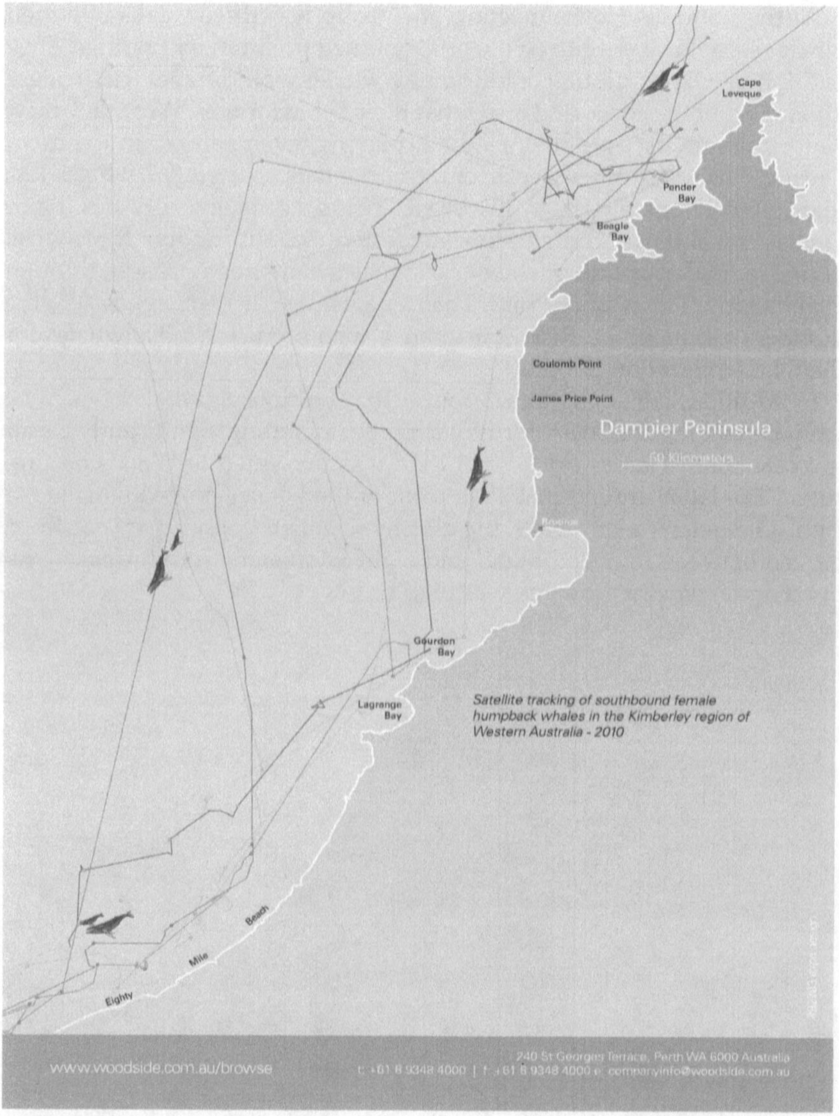

Figure 6.6. Whale Migration Path as Projected by Woodside Consultants. Photo: Carsten Wergin.

The reason why citizen scientists found significantly more whales inside the eight km range is the fact that 80 percent of the studies for government and industry were conducted at low tide. Thus, like the report on bilbies, they were based on a biased research model that was informed by Western scientific standards alone and disregarded the Indigenous knowledge available about the area. The EPA was led to conclude that Walmadany/James Price Point would not be important to vital whale activities such as breeding, resting, and playing.

The numerous resting and playing behaviors observed by citizen scientists proved the opposite. Their findings were made public in March 2013, when a delegation of citizen scientists traveled to Canberra and met with the Office of the Federal Minister of the Environment to deliver their survey. During their three-month study, they had counted a total of 3,343 individual humpback whales, including 214 cow-calf pairs. Taking a conservative approach and removing potential re-sightings, this means that they recorded at least 2,669 individual humpback whales, including 172 cow-calf pairs.

The results suggest that between 12,108 and 15,876 whales passed within eight km of Walmadany/James Price Point during the 2012 migration season (Goolarabooloo et al. 2013). Meanwhile, both the conservative estimate of 2,669 whales and the absolute maximum of 15,876 whales greatly exceeded the 1,000 whales estimated by RPS Consultants. The failure of the government and industry consultants to consider tidal movements while conducting their surveys contributed to this significant underestimation.

Adapting their findings to Western scientific terminology, citizen scientists concluded that the proposed LNG precinct would cause significant habitat degradation and behavioral disturbance to humpback whales due to acoustic pollution, possible ship strikes, changing water quality, and turbidity from ongoing dredging and shipping movements. This, in turn, was likely to cause the exclusion of whales from appropriate habitat and reduce their reproductive success.

In light of the above, citizen scientists again proved the science that informed the EPA decision to be inadequate and not comprehensive, stemming from poor temporal and spatial sampling and often relying on inadequate methods (Goolarabooloo et al. 2013). However, despite their efforts, on 18 December 2012 the EPA approved the LNG development at Walmadany/James Price Point. Its chairman, Paul Vogel—who was also the sole decision-maker in this process, as the four other board members had by then withdrawn from it due to conflicts of interest (some were Woodside shareholders)—announced in his statement:

> The public comment period was an opportunity for the public to outline whether this proposal should or should not be declared a derived proposal and their reasons to support their view (. . .) Following rigorous evaluation of Woodside's proposal, no new environmental issues were raised that had not been adequately addressed in the strategic proposal nor was there any significant new or additional information that justifies reassessment. (EPA 2012, n.p.)

Subsequently, in August 2013, the Supreme Court of WA ruled the EPA's assessment process invalid due to the board members' aforementioned conflicts of interest. The results of the citizen science projects did not have any official impact on this decision. However, on 26 November 2014, *ABC News* reported:

> Following the [WA] State Government's decision to proceed with the approvals process, three independent scientists have been brought in to complete a new assessment. The trio, appointed by the Environment Minister in February, is being headed up by environmental consultant Gerard Early. Mr Early said people can have faith in the process. "All three of the assessors are completely independent, so it's a completely separate exercise," he said. "We've deliberately not looked at the original EPA report." The delegates are spending two days at James Price Point and in Broome, inspecting the site and discussing the proposal with the local shire, conservation groups and traditional owners. (ABC News 2014)

While this initiative refrained from drawing on the previous research conducted, its delegates again spent only a very short period in Broome and at Walmadany/James Price Point—namely two days. The announcement of consultations with conservation groups and TOs in the quote above gave hope that this time around more attention would be paid to the results of the citizen science projects. However, in its decision announced on 26 October 2015 the EPA again approved an LNG development at Walmadany/James Price Point, if a list of "28 strict conditions" was fulfilled. The public was once more invited to submit appeals by 9 November 2015. To the best of my knowledge, at the time of writing, a final decision from the WA Minister for the Environment is still pending.

6.3 CARE AND CONCERN

In the assessment of the environmental impact of the proposed development at Walmadany/James Price Point, the proponents of development dealt with uncertainties as inevitable factors, but with a strong aim that these uncertainties be downplayed, ignored, or eliminated so that they would not disturb the "fair go" for economic opportunities to unfold. In other words, supposedly impartial science was instrumentalized for the pursuit of economic gain.

This was unsuccessful due to the citizen science projects, the results of which deconstructed government and industry rhetoric of scientific rigor and accuracy. However, the same findings were largely dismissed by government authorities, much like the Indigenous knowledge that contributed to them. This is accounted for by the fact that a Settler-colonist view is still advocated by Australian government and law, to the extent that neither is blind or impartial. Instead, both override the transecology they are meant to preside over.

In opposition to this, citizen science was conducted with the help of diverse people—academics, lawyers, media representatives, international tourists, and travelers—who dedicated their time and energy to the cause *pro bono*. Together with their Indigenous collaborators, they chose to defeat Western science on its own quantifiable terms: among themselves, they recognized the help of "living country," of living heritage, and of Indigenous knowledge, but in their reports they concentrated on using Western scientific jargon. Anything else—that is, the acknowledgment of "living country," let alone recognition such as the *Goolarabooloo*'s that "country put [bilbies] there"—would have offered an opportunity for the proponents to undermine their authority as "irrational."

However, instead of giving in to the "opportunity dreaming" advocated by government and industry, the citizen science projects show a means of generating sound research results from active engagement in transecological entanglements. Producing environmental assessment reports therefore turns out to be more than a necessity. Such reports become biocultural projects that take Indigenous knowledge seriously, while the ecology they describe becomes a "regime of justification" in its own right, rather than being submitted to other regimes such as Settler-colonialist representationism or mainstream economics.

Louis Dumont's (1986) *Essays on Individualism* are among the first anthropological texts to engage in the establishment of such regimes. They describe how individual salvation was translated from religiosity into state and market economics: whereas the retreat into a monastery was meant to facilitate a search for an inner self in solitude and lead individuals toward outworldly salvation, market economics promised to achieve this through the fulfillment of supposed innerworldly desires: "We have here the model of modern artificialism at large, the systematic application to the things of this world of an extrinsic, imposed value. Not a value derived from our belonging in this world, such as its harmony or our harmony with it, but a value rooted in our heterogeneity in relation to it" (Dumont 1986, 56). Rather than emphasize what people do in the world, market economics advocates aspirations for what might (never) be, attached to the promise of the opportunity of a better life—"a fair go." Such a world(view) full of aspirations sees people dream of opportunities rather than face realities. What the ethnographic data have shown is how the *Lurujarri Heritage Trail*, and with it the world of the *Goolarabooloo*, hinders this opportunity dreaming from literally taking place.

In my final ethnographic chapter, I draw on a personal account of how the impossibility of proper translation/representation of Indigenous knowledge, Lore, and culture in Western scientific terms highlights the multiple ontologies at play in the pursuit to protect Walmadany/James Price Point from industrial development. It therefore turns to another aspect of the transecological experience itself and how it opens up a different perspective on what counts as good, what is worth preserving, and on what grounds.

NOTES

1. Blackbirding refers to the enslavement of South Pacific islanders on the cotton and sugar plantations of Queensland, Australia, as much as that of Australian Indigenous people by the pearling industry in Western Australia. The practice was especially prevalent between 1847 and 1904. Meanwhile, in the sense that it involved the coercion of people through trickery and kidnapping, blackbirding continues to this day, for example, with regard to Indigenous people in Central America who are forced at gunpoint to work as plantation laborers.

2. This differentiates the Dampier Peninsula from the desert context, where succession is part of the system because it is not a system based on descent only: in the rigorous environment of the Western Desert, there is a need to accommodate people in various ways in order to make a system survive.

3. The following dialogue is derived from my handwritten notes and thus is not to be mistaken for a proper transcript of the exchanges that took place. I have opted to present them in this way for stylistic reasons and because I nevertheless feel confident that I have recorded their content appropriately. Any mistakes are attributable only to myself.

4. In contrast to this, industry consultants only briefly visited Broome and took findings back to their home institutions to be processed. The latter would only reappear in Broome as graphs, in PowerPoint presentations, or on paper in letterboxes, where they were used to downplay the transecological dimensions of Walmadany/James Price Point (figure 6.6).

5. At the Lotteries House meeting, citizen scientists pointed out that such technology is quite affordable, as some cameras "are available at Bunnings [Australian warehouse chain] for only about $200 AUS Dollar." A lecturer from La Trobe University Melbourne provided additional cameras. Citizen scientists explained that he had participated in the *Lurujarri Heritage Trail* for many years and saw this as a way to contribute to the opposition movement.

6. Video footage of the bilbies discovered by this citizen science project is accessible via the URL: https://www.youtube.com/watch?v=kOpAWcsX_hs (Accessed July 1, 2023).

7. Instead, proponents downplayed the environmental impact of the LNG precinct. Another example of this that citizen scientists put forward that evening at Lotteries House was that the aforementioned report acknowledged that the Walmadany/James Price Point coastline accounts for more baitfish than the Great Barrier Reef. However, the summary of the same report only mentioned the large amount of fish. It left out the comparison to the Great Barrier Reef, the iconic site

with UNESCO World Heritage status, mention of which would have given significantly more impetus to the statement.

8. Consequently, the EPA argued, the impact of the LNG precinct would not result in a great loss. This conclusion of a potentially larger habitat was again based on false assumptions. The semi-nomadic nature of bilbies was assumed to mean that these animals could easily move to new areas to cope with a loss of habitat. However, bilbies are semi-nomadic only if there is a need for them to compensate for low food resources. Consequently, a loss of prime habitat such as Walmadany/James Price Point could well negatively impact their general migration behavior (The Wilderness Society et al. 2012).

9. Figure 6.6 is taken from one of the aforementioned handouts that were posted to the wider Broome community. It makes a particular case in point, as the image suggests that whales literally "avoid" Walmadany/James Price Point and even without any LNG facility built.

Chapter 7

Collaborative Science

As an anthropologist, I am trained to acquire data using a variety of methods and to contribute to scholarly discourse through ethnographic writing. There are a significant number of anthropologists and anthropological institutes throughout Australia whose academic staff work in very much this way. However, during fieldwork I was equally impressed by the demand from government and industry for anthropologists to work as consultants and expert witnesses. These anthropologists have the difficult task of using their qualifications as a means to assess an Indigenous person and their extended family status with regard to, for example, a claim made for a particular stretch of land. These assessments are also used to establish whether a site that is destined for resource exploration or other industrialization proposals, like Walmadany/James Price Point, has sufficient cultural significance to prevent development. Anthropologists present the evidence that Indigenous and non-Indigenous people give them in lay language in their scientific reports, which in turn assist in the assessment processes of Native Title claims, resource exploration proposals, and other activities with possible impacts on country. With such evidence at hand, decision-makers within a government or courts of law speak about the environment, assess its material and cultural values, and ultimately determine its future.

I had introduced myself to the *Goolarabooloo* as an anthropologist. This for me meant that I had relevant academic training. What they understood from it, however, was that I was entitled to function as an expert in the ways just described. I was therefore asked to write a scientific report about the cultural and heritage values of Walmadany/James Prices Point.

This was a new field for me, but I was motivated to give back to the community that had agreed to host my research project, so I did not hesitate when the *Goolarabooloo* asked me to produce such a report about diverse heritage sites in the vicinity of the proposed LNG facility.

After my immediate agreement, however, I had second thoughts, because I was not familiar with the writing style of reports that are used as evidence in a court of law. Thus, I was uncertain as to whether my contribution would have the desired impact. I contacted my "scientist in charge" on how to proceed, and together we developed a toolkit that I took along for the survey I conducted "on country." Apart from the usual camera, voice recorder, notebook, and pen that anthropologists are commonly equipped with, this also included a GPS device and compass.

In January 2013, in the middle of the wet season where temperatures and humidity are extremely high, I found myself walking for two days across the extended dune system near Walmadany/James Price Point, together with Phil Roe and a small group from the very inner circle of *Goolarabooloo* supporters. We chose the mornings, when the temperature was still bearable, to document burial grounds, artifacts, and other cultural remains.

While I usually include my personal experience in the documentation of such a research endeavor, I opted to stay as matter-of-fact as possible regarding what I was presented with and told. Having attended previous meetings of No Gas campaigners where they gave advice on similar matters, I saw a danger that the report would fail if it was termed subjective or unscientific. Instead of personal accounts, I therefore drew on ethnographic and archaeological techniques as well as Western scientific terminology—or, in the language of my Indigenous hosts, "white magic"—to translate the Indigenous world(view) that was presented to me into scientific evidence (figure 7.1).

I encountered a problem with this process of translation, because what I wrote down as matter-of-factly as possible seemed to lead to the subjection, even loss, of this world(view) for the benefit of the scientific one. In the remainder of this chapter, I am therefore less concerned with the question of how the world(view) of the *Goolarabooloo* is best evidenced. Instead, I draw on the experience I had while preparing my report to highlight the Modernist constraints that I found necessary to inflict on the "living country" and living heritage I wrote about. This ethnographic chapter thus returns to my general motivation in this project, which is less about producing an ethnographic account of an Indigenous world(view) under Settler-colonialist threat, and more about the Settler-colonialist world(view) itself and how it operates and manifests in contemporary Australia, actively shapes the future of people and places, and also its potential alternatives.

Figure 7.1. My Survey Toolkit. Photo: Carsten Wergin.

In order to fulfill the expectations of the *Goolarabooloo* and their lawyers and provide a report that qualified as such in a court of law, I had to employ "white magic"—that is, to translate the heritage values and other qualities of "living country" into Western terminology. Relevant members of the protest movement and wider Broome community still term my report a success. In 2017, a senior figure in the movement told me in a personal conversation that it "really frightened" some of the stakeholders in the LNG proposal when it was presented to them in a meeting he attended.

Why it frightened them remains a mystery. Because of the burial sites I mentioned and the possible haunting by disturbed spirits? Or because of the negative press it would provoke for the stakeholders if an LNG facility were to be built on top of those? Maybe both. I am not concerned here with the success of the report. On the contrary, as stated above, what motivated this chapter was how I found my attempt to translate parts of an Indigenous world(view) into the terminologies of Western science and law a failure.

Bugarrigarra recognizes the entanglement of people and place. But translating this into Western science terminology turned out to be not a "last frontier," as tourist advertisements for the Kimberley region like to have it, but a "lasting" one. My report had the desired effect, as it added recognized evidence for why to oppose the industrialization project. However, from a transecological perspective it failed because it could only represent what Walmadany/James Price Point stands for as a cultural site. It could not substitute for the lived experience that is central to the *Lurujarri Heritage Trail*.

My report was written with the sole purpose of meeting the expectations spelled out in heritage laws and other formats, not to challenge their integrity. With the success of this translation of the Indigenous world(view), the work contributes to the "mainstreaming" of *Bugarrigarra* into Western law and science, which is necessary because of the general weakness of the latter in acknowledging different ways of being in the world. On the one hand, my report contributed to filling a general lack. On the other, it highlighted that different means are needed in order to properly account for transecological entanglements, rather than substituting these with scientific or lawful representations of reality. In conclusion, I argue that an applied anthropology that emphasizes transecological qualities of processuality, practice, and experience over representationism can play a crucial role in this quest.

7.1 OVERCOMING WESTERN MYTHOLOGIES IN TRANSECOLOGICAL REALITY

In any conflict situation centered on industrialization versus preservation, Aboriginal people find themselves in a dilemma: They need to prove, for example in courts of law, that their connection to the land is alive and that they have the right to speak for it. However, they cannot do so in terms of Aboriginal Lore and culture because of the need to adapt to *whitefella* law, which is "dominated by a western legal demand for certainty, based on a *legalistic ontology of determinacy*" (Smith 2007, 125, emphasis added). Along these lines, Holbraad, Pedersen, and Viveiros de Castro (2014, n.p.) state that "one does not need much anthropology to join the struggle against the political domination and economic exploitation of Indigenous peoples across the world."

In my case, "joining the struggle" entailed more than a commitment to support the rights of Indigenous people: It meant questioning the very discipline of anthropology I was trained in. I came to realize this firsthand when I was recruited by the *Goolarabooloo* to document sites for their scientific report. The problem that unfolded during this process was that a proper translation of the significance of "living country" and its living heritage, true for example with regard to the continued use of sacred sites, into the terminology of Western law and science was impossible. What I encountered was too different from the format I was required to account for it in. This is why, as stated above, the request for me to produce a report became an exploration of the powers of Settler-colonialist discourse.

Documentation of significant aspects of Aboriginal Lore and culture in Western scientific terminology is nothing new. In the Kimberley, it notably ties in with a great deal of historic work conducted by German anthropologists and missionaries, as well as their particular ways of accounting for ontological differences. One prominent example is the anthropologist Helmut Petri. In 1938, he led the first Frobenius expedition from Germany to the Kimberley. "Frobenius XXI" marked the beginning of an extensive relationship between Petri, his wife Gisela Petri-Odermann, and the people and region that lasted until his death in 1986.

In his article "Sogenannte 'schamanistische' Erscheinungsformen im heutigen Nordwest-Australien" (1987, "So-called 'shamanistic' manifestations in today's Northwest Australia") Petri finds a remarkable way to account for one aspect of the ontological differences he encountered. He remembers a conversation he had in 1938 at Munja Station in the northern Kimberley (Petri 1987, 614). He spoke about kinship with a person named Yáobida (the "speaker" or "gossiper"), while his colleague D. C. Fox had left for Brockmann Creek some 50 kilometers away. Upon his return the two researchers shared their activities of the day and realized that both of them had spoken with Yáobida more or less simultaneously in those two distant places. Yáobida was, Petri explains, a healer (*bánman*, or *mában*). He was equipped with the ability to materialize his alter ego and could thus be present in two different places at once.

Petri's treatment of Yáobida and his abilities was mind-opening with regard to my own engagement with the world of the *Goolarabooloo* and my documentation of its living heritage. Instead of questioning Yáobida, Petri writes that "[wir müssen] es so hinnehmen, wie es ist" (Petri 1987, 614, "we have to accept it, as it is"). Petri does not present a rational explanation for the events but instead acknowledges their possibility. He neither represents the events nor draws on Western representationism to explain and in the process alter the lived experience. Instead, he remains within the realms of the experience that he and Fox shared that day. He emphasizes the intra-actions between himself, Yáobida, and Cox, and thus gives a truly ethnographic account.

This is in contrast to the work of another German-speaking anthropologist in the Kimberley, Erich Kolig. Like Petri, his views stem from long-term fieldwork experience. However, Yáobida's abilities and Petri's account of them would have fallen under Kolig's concept of "myth-consciousness," which he defines as a central problem for Aboriginal people who need to assimilate with/in Settler-colonialist Australia. "Myth-consciousness," he writes:

> reduces the flux of time and events, of people and their deeds, to epiphenomenality and hence to relative triviality. This proclivity obviates the need to be curious about the non-mythical or non-creative, "historical" past, or about the future. It also eliminates the need for carefully monitoring the present in

order to elicit trends and processes that might be projected into the future, and the need to be able to anticipate it to some, albeit minor, extent. (Kolig 2000, 10–11)

In other words, for Kolig, the "myth-consciousness" of Aboriginal people deprives them of the ability to think ahead and envision their future within Australian Settler-colonialist society. However, Kolig did not consider the possibility of a "myth-consciousness" parallel to Settler-colonialist reality. Paddy Roe's notion of sending out and walking on his dream, the *Lurujarri Heritage Trail*, is a case in point. Frans Hoogland describes this with regard to the continued presence of Aboriginal cosmology in "living country" and heritage, as follows:

The other story of the birth of this Song Cycle is that the spirit took a body and then started the sounds [claps his hands], and started the dance [claps his hands], and then the song, then sending out a dream relating their landscape and then walking on it. Same like Old Man [Paddy Roe] said: He sent out, they start walking on it [the *Lurujarri Heritage Trail*]. So the guy who had a Dreaming . . . is always here (. . .) Everything is always now (. . .) We are part of the past, part of the future, and part of the present. It's an ongoing process (. . .) It's always evolving (. . .) This culture, us human, are part of a greater process. You're not a subject to an external greater process and you kind of walk in there. You are a part of it. You are an imprint in the process.

That's the land, too (. . .) If you think about the Song Cycle system, it is all different areas for different purposes: This one to make more mosquitoes or more babies, and this one to camp and rest, and this one is for dreaming (. . .) The reality is always subject to the great spirituality of it (. . .) It is being part. . . . Being part of the ground you're walking on. If you make a boomerang out of a tree, that tree's function always has to be for that boomerang. That has been its job: To make that thing out of it. You have to respect that tree because it is going to give you life. You don't kill that tree. You cut an arm of that tree. You cut a part of him. He'll stay alive. You don't kill him. Out of that part, you make sound. And the sound creates life (. . .) And the other tree you use, it's a different thing, and it is also to create life. So, it comes back to sound again. It comes back to dreaming again. It is not fanciful dreaming. It is "manifesting-the-reality-dreaming" (. . .) It is very hard now to separate that when you got one heavy materialistic dream of the things of necessity we all use and need today: "We should support that materialistic thing in order to maintain this reality we live in." But on another level, we destroy the mother. (Transcript 120807_4_FH)

Hoogland criticizes what I have described as opportunity dreaming (Wergin 2016)—that is, the materialistic dream that departs from the "manifesting-the-reality-dreaming," the here and now. This spiritual departure from what is in place, which is expected of Australia's Indigenous population, is one of the four pillars on which its Settler-colonialism rests. The opportunity dreaming is a biopolitical tool for restructuring people's daily lives according to a Western model that is based on ideals of development and progress. Or, in Kolig's terms, it is the "myth-consciousness" of the West, but with more dramatic consequences for people and environments than Aboriginal myth-consciousness can be made accountable for. Because Aboriginal myth-consciousness is inclusive, and embraces the past, the present, and the future, while the opportunity dreaming is exclusive and asks Aboriginal people to forget their past and ascribe to the Settler-colonialist values of the present, without which they would supposedly have no future.

In sum, Petri and Kolig, two of the most prominent German-speaking anthropologists in the Kimberley, present two opposing ends of anthropological engagement with Indigenous Australia: one acknowledges myths as entangled with reality (Petri); the other sees the need to overcome myths in the face of Settler-colonialist reality (Kolig). My observations support Petri's position, in which he does not subsume his experience with Yáobida to prevailing Western thought. His view is the one that allows for the recognition of both multiple world(view)s and the ability of (Indigenous) people to consciously move between them. What is equally inspiring for me is how Petri's rather simple statement that it is necessary to "accept things as they are" comes across without him losing his integrity as a researcher. This is due to the fact that he is immersed in the field. His writing speaks the language of experience.

During my own fieldwork, Kimberley researchers involved in the *Karajarri* Native Title determination presented a further case in point when they told me about interviews they conducted with Indigenous Elders. Some of those remembered how Petri examined them in their childhood. The West Australian government had granted him the status of "Protector of Natives," which entitled him to examine children's bodies for signs of leprosy. Based on his judgment, children were sent to camps set up to accommodate those infected.

This anecdote is worth noting as it gives a remarkable twist to the title Petri chose for his 1954 publication in German of *The Dying World in Northwest Australia* (Petri 2011), in which he discussed some results of the 1938–9 "Frobenius XXI" expedition. It is not known whether Petri chose the title to announce the imminent loss of Aboriginal culture and heritage in the region due to the impact of Settler-colonialism, or whether he meant to commemorate his experiences of examining children for signs of leprosy. The mere possibility of the latter confirms that there is always more than one myth-consciousness at play, and that it is a quality of the ethnographer to accommodate both: title and anecdote once again show the close entanglement of Petri's ethnographic experience and his writing.

In the following section, I present a particular ethnographic moment to describe these qualities of entangled ethnographic experience and writing in more detail. In it, I continue to reflect on my exercise of what my Indigenous collaborators ironically termed "white magic," which is applied when Indigenous knowledge is translated into Western scientific terminology. I draw on the fundamental problem of representationism that I outlined previously, which becomes most apparent in these situations, in order to argue for what lies beneath: the transecology of Walmadany/James Price Point, a living heritage that is both immanent and beyond translatability.

7.2 "WHITE MAGIC"

Construction of the LNG facility was supposed to begin in 2009. The plant was meant to process and export natural gas extracted from the Browse Basin in the Indian Ocean, roughly 425 kilometers offshore. The diverse range of opponents successfully delayed the project until April 2013, when Woodside withdrew from it, officially for economic reasons. Meanwhile, for the previous five years and more, the protest had created such a delay to the construction work that it can equally be argued that this is what actually prevented the multibillion AUD project from literally *taking place*.

When I visited Walmadany/James Price Point for the first time in 2012, it had turned from an overnight stop for Trail participants into a fully functioning protest camp with at times more than thirty residents. Solar panels had been installed to supply energy. Long antennas were visible from afar and connected phones, radios, and computers with the rest of the world. A wooden tower had been erected as a lookout for activity on the dune system. There had been various unannounced visits by workers from Woodside, accompanied by security personnel from Hostile Environment Services, government officials, and scientists, and protesters were vigilant, because "living country" could be affected any time, be it by taking soil samples, or setting up drill rigs.

A banner with the words "First they ignore you, then they laugh at you, then they fight you, then you win" hung at the entrance to the protest camp (figure 7.2).

This famous quote by Gandhi captures the conflict in various ways. If "you" describes those who dwell in the camp, "they" might address the political and industry groups that planned to build the LNG facility. However, "they" might also refer to Aboriginal people with family ties in the region, who were once forcibly removed as part of the Stolen Generations or had for other reasons moved away, but who now had come back

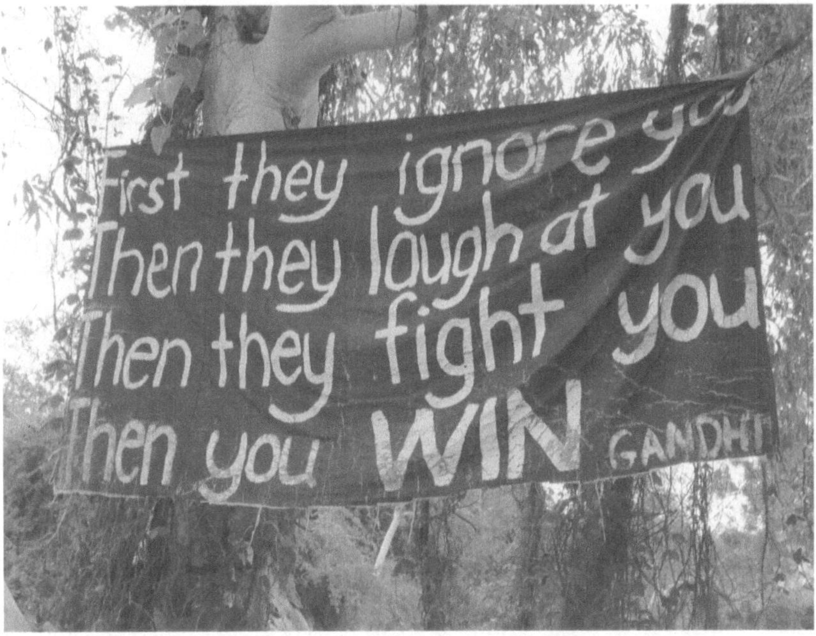

Figure 7.2. Banner Near the Entrance to Protest Camp Walmadany. Photo: Carsten Wergin.

to claim the land. Opponents of the industrialization project described some of the latter as people who are not called by "living country" to protect its heritage, but who hear "money calling," such as the $1.5 billion AUS Dollar compensation package attached to the original industrialization proposal.

Yet again, "they" could also be understood to describe the scientists and environmentalists active on the land on behalf of either of the conflicting groups, who might consider their ways of acquiring and representing knowledge superior to Indigenous Lore and culture—and working for different interest groups even led some scientific experts to question each other's authority, because ethnographic studies commissioned by government and industry could either suit the purpose of Indigenous "heritage clearance," or end up in a filing cabinet, never to be published. Rights to such studies do not necessarily rest with the Indigenous groups they speak about, but with those who pay for them to be produced instead.

As outlined above, my involvement in the conflict over the industrialization of Walmadany/James Price Point was different. Neither industry nor government was interested in my work, let alone paying for it. My training as an anthropologist meant that I was equally qualified to produce scientific reports. However, I had not come to defend the qualities and heritage values of this "living country." Originally, my research aimed to address the Trail's qualities as an Indigenous tourism experience.

Be that as it may, the *Goolarabooloo* thought that my anthropological work could make a contribution to their opposition movement. I brought with me academic degrees and anthropological expertise, scientific tools, language skills, official EU logos, and an air of impartiality. The contribution I eventually made ties in with those of other supporters who came from a range of professions, and who found a common ground on which to accommodate their diverse world(view)s: the recognition of "living country."

One particularly media-sensitive event that showcased the related "nowhere else but here" mentality described above, which was shared by these collaborators, was the arrival of the *Steve Irwin* (figure 7.3). This ship, belonging to the environmentalist group Sea Shepherd, came to Broome and the Kimberley coast in August 2012 as part of *Operation Kimberley Miinimbi* to raise awareness of the humpback whale population that migrates along the coast of Walmadany/James Price Point each year, and that would be impacted by the LNG processing hub. The visit had support from both the Australian Labor Party, which was represented by Melissa Parke, then representative for the Division of Fremantle, and the Australian Greens, represented by their former leader and senator Dr Bob

Figure 7.3. Welcome Event for the *Steve Irwin* and Its Crew. Photo: Carsten Wergin.

Brown. Both politicians came to Broome for the occasion and attracted additional media attention.

The protection of whales marks a point of departure for many well-known environmental groups such as Sea Shepherd and Greenpeace. In the 1970s, the latter rose to fame when it set out from San Francisco to intercept Soviet whaling ships in the Pacific. Since then, endangered whales in their calving grounds have been a significant drawcard to create national media attention, which it did for the conflict over Walmadany/James Price Point. The possibility of observing migratory humpback whales from a very close distance is well-known and has traditionally attracted many tourists to the region.

Taking a transecological position that emphasizes entanglements and collaboration, the whales can be termed active participants in this media-sensitive event at Gantheaume Point in Broome. Their presence provided a valid reason for collaborators to join forces and added another sensitive dimension to the story about the possible industrialization of Walmadany/James Price Point, which generated further awareness far beyond the Kimberley. To draw on the above quote by Gandhi, after ignorance and laughter, industry and government started having to acknowledge and ultimately confront/fight an intensifying protest whose transecological forces made arguments not only heard but also felt on a national level, not least due to the emotions evoked by imaginaries of endangered whales and their calves.

In other words, "living country" and its living heritage were able to draw powerful human and more-than-human protesters together—and not for the first time. I was able to begin my report with reference to numerous previous studies published about the area, on which Indigenous custodians and scientific experts collaborated. One of the most comprehensive such studies to date is the aforementioned *Management Report for the Lurujarri Heritage Trail*, written by Elizabeth Bradshaw and Rachel Fry (1989) and commissioned by the Western Australian Museum. Walmadany is named as an important *jila* (waterhole) in this study (figure 7.4). Aboriginal people used the area around it as campgrounds (Akerman 1981). As a result:

> Burials are periodically exposed and re-covered by shifting sands but the locations of only a few of these are known [. . .] According to the traditional custodian [Paddy Roe] it was customary to bury the dead in shallow graves in the dunes along the coast. It is to be expected that any developments in the area would disturb the bones of the many people who are buried in the sand hills between Bindingankuny and Minyirr.
>
> (Bradshaw and Fry 1989, 9)

Figure 7.4. The Jila (Waterhole) at Walmadany/James Price Point. Photo: Carsten Wergin.

Walmadany and its dunes are furthermore defined as, "A MAJOR [*sic*] site. It has archaeological integrity and dense material over extensive areas (i.e., 10,000+ pieces) of shell/bone/stone material and some of the following features: hearths, species-specific scatters, bone remains, an association with older soils" (Bradshaw and Fry 1989, 12). Bradshaw and Fry describe Walmadany as a space of 2,500 x 500 meters. The area shows "middens" and "artefacts," was used as a "camping area," and is "mythological" and of "major archaeological and ethnographic significance, plentiful material over extensive area on and in intact lenses within cliff face and in dune blowouts." The authors conclude that it should be the northern tip of a "proposed Lurujarri Protected Area" (ibid., 14).

This 1989 report by Bradshaw and Fry had an impact on many of those involved in the protest for Walmadany/James Price Point, notably EK's director Martin Pritchard:

> When I read [the report], I realized that this is an incredible bit of coastline. When you look at all the documented sites and description of those sites, and the fact that not all the sites were in there because there were the sites for initiated people only to know about. . . . There was 60 sites that were named in an 80-kilometres stretch. Some of them one-and-a-half kilometres long.

> You're basically looking at an unbroken series of cultural and archaeological sites along this piece of coast. When I got to understand that [Mr Roe] was the senior law boss and that Richard [Hunter] was the man with the stories coming back from ancestors (. . .) since time began according to Aboriginal people. . . . Then Phillip had responsibility for the songs. It's just like I was blown away! It made me just scratch my head even more. Why would you want to destroy something like this? (. . .) That made me realize just how even more special it was… on top of the environmental importance of it. . . . [It] gave me a deeper insight into Aboriginal culture and the fact that it's still living. I (. . .) could meet people and talk to people that were still carrying out their traditions regardless of colonization and industrialization. I'm actually just being able to go out there myself and through those dunes and find artefacts sitting there: Stones, grinding stones, baler shells, middens. . . . It was (. . .) just amazing.
>
> What astounded me was that, this is probably one of the most well documented large areas of Aboriginal land in Australia and the connections. Two books being written on it, by Stephen Muecke, another one's in the making. Paddy Roe been in numerous films and documentaries and BBC having been here years ago as part of the *Millennium Series* and all kind of stuff. It was quite perplexing that they'd chosen this particular site when they should've known how significant it was and how well documented it was. I think there was either incompetence in not knowing that, or it was deliberately held back. Yes, I found that a very disturbing part of the whole thing really. (Transcript 130513_1_MP)

I was no less surprised than Martin Pritchard at how a place worth protecting in this way could, less than twenty years after the publication of a scientific report on its heritage values, be proposed as a site for large-scale industrialization. Had no one read the reports? Was it possible to deliberately ignore them? After all, they were commissioned by a public institution, and as such were produced with government funding. In my report, I therefore opted to both emphasize the existing reports and add a contemporary dimension to them based on my own material. This allowed me to reference Bradshaw and Fry, and to remind potential readers of their work—just in case.

In late January 2013, Phillip Roe guided me through a section of the dune system, similar to the way his grandfather Paddy Roe had guided Bradshaw and Fry twenty-five years before. He pointed out burial sites, artifacts, bones, and campgrounds that had recently been exposed by rain and wind. His explanations were complemented by data I had collected from other studies, and during my participation in the *Lurujarri Heritage Trail* in July 2012.

Following strong winds and rain, Roe patrols the area to maintain sites and rebury exposed objects and human remains. During our visit, he traced the lives of people, identified communal areas rich with middens, and pointed out artifacts and campsites. Our walk began at a place he identified as the burial site of Walmadany, the Aboriginal warrior who gave the area its name.

About 15 meters away, near an exposed turtle bone, Roe pointed to another burial site, this time of an Aboriginal Elder, or what he described as an "old fella":

> *Phillip Roe:* That was uncovered about two years ago, but we buried it, more deep, covered him properly again. . . Even this one here.
>
> *Carsten Wergin:* So when did you find that other bone? Last week only?
>
> *PR:* Yeah. I buried it up, and usually, if we would have no rain and wind, you could have seen where it lay, uncovered.
>
> *CW:* And how did the turtle bone get up here (figure 7.6)?
>
> *PR:* They carried it man, from down there . . . With that fireplace up there, where they must have been cooking (figure 7.5). And this is where they must have ate. Must have been these old people sitting here, and this is where this old fella must have died.
>
> *CW:* While eating the turtle?

Figure 7.5. Fireplace. Photo: Carsten Wergin.

PR: I'm not sure.

CW: And how would you use [this grinding stone] (figures 7.7/7.8)?

PR: Well, you just grind it, grind seeds and things. And those holes in the middle, that's when they used to hit shells.

Figure 7.6. Close Up of Turtle Bone. Photo: Carsten Wergin.

Figure 7.7. Phillip Roe Using the Grinding Stone. Photo: Carsten Wergin.

Figure 7.8. Phillip Roe Using the Grinding Stone. Photo: Carsten Wergin.

In contrast to archeological uncovering of sites through the scientific technique of excavation, Roe and I visited sites that are regularly uncovered by wind and rain. In other words, they are entangled with the transecology of the 'living country' and expressions of its living heritage. Wind and rain can thus be termed additional more-than-human collaborators, including in the research I undertook. Their active participation impacted it, continuously reshaping the transecology I was able to engage in.

As the whales assisted in the sensitive staging of the arrival of the *Steve Irwin*, which generated further environmental and political support, wind and rain assisted "living country" in providing the information that I needed to make a convincing argument in my report. Because of the lived experience I was provided with, Roe did not need to represent the significance of the place to me through story or other imaginaries. During our survey, it presented itself to us.

The cultural significance of Walmadany/James Price Point is not accessible through mere knowledge about where something or someone is buried, let alone the scientific expertise to document this in Western terms. What is needed is the knowledge, not only to identify what is exposed, but also how to rebury it—how to actively engage in the transecology of which such living heritage forms part and provides access to. Roe and other custodians do not act on "living country" and living heritage, but act together with it, through (re)actions that make a prime example of their entanglement with it. As Petri said of Yáobida's abilities, "we have to accept it as it is."

Much like Petri, I would have literally liked to leave it at that. However, I had committed myself to converting my observations on country into Western scientific terminology. The tools I had brought for this occasion (GPS logger, tape measure, camera, and voice recorder) allowed my report to include facts about geographical position, size, utility, and other quantifications. For example, I stated that the number of sites that Roe identified indicated that the dunes around Walmadany/James Price Point had been extensively used by Aboriginal people. The land supplied sufficient food, water and shelter, making it possible for them to live in this area year-round (figure 7.9).

Again in line with what Bradshaw and Fry had done twenty-five years before, I came to the conclusion that the dune formations must be considered one interrelated network of campgrounds, burial sites, and mythological places. The knowledge about this network has been passed on through the generations; clearly, I concluded, Walmadany—both the place and the spirit/person—remain of importance to the lives of Indigenous people who associate themselves with the area. What I did not mention was the transecological dimension this "importance" entails, which is explicable only through the experience of being on "living country," through (re)actions caused by the interferences of wind and rain, or

Figure 7.9. View South Overlooking the Dunes System. Photo: Carsten Wergin.

a dolphin in a creek that reminds me to fill up my water supply before going any further.

After reviewing my report, despite the attempts I had made to translate transecology into scientific terminology, lawyers who worked for the *Goolarabooloo* pushed me further into representationism. They recommended speaking as clearly as possible in the terms used to describe the law my report was to address. I thus paraphrased Section 5 of the Aboriginal Heritage Act 1972 (WA). The Act states that the importance of a cultural site is acknowledged if:

(a) the area holds numerous places of importance and significance where persons of Aboriginal descent have left objects used for, as well as made or adapted objects for use for, purposes connected with the traditional cultural life of Aboriginal people, past and present;
(b) it includes sacred, ritual and ceremonial sites, which are of importance and special significance to persons of Aboriginal descent;
(c) it continues to be associated with local Aboriginal people and is of great historical, anthropological, archaeological and ethnographical interest and should be preserved because of its importance and significance to the cultural heritage of the State of WA and the Australian nation.

In my descriptions, I made use of direct quotation from this act, or language as close as possible to the original wording, in order to meet the requirements set out by law. In the conclusion of my report, I also referred to Section 6 of the Aboriginal Heritage Act 1972 (WA), which addresses the "continued use of a site." This might have been the most suitable law, as the continued use of "living country" was evident in the fact that the preservation of its living heritage, including both artifacts and human remains, remained part of daily activity for the *Goolarabooloo*, who continued to patrol the land and present the public aspects of it to participants in the *Lurujarri Heritage Trail*.

Meanwhile, this also required me to translate Roe's explanation, "we buried it, more deep, covered him properly again" into the technical term "continued use of a site." The *Goolarabooloo* called this process "white magic," but not in the conventional sense of the term with its positive connotations of supernatural powers used for selfless purposes. Clearly, neither of us was enthusiastic about "white magic," because it meant that the transecological component that makes up the significance of the place had to be hidden, overridden—whitewashed.

While the *Goolarabooloo* enjoy Walmadany/James Price Point, take their family out hunting and fishing, and have means to cater for its significance in its own terms, through lived experience and what Barad terms intra-action with it, they were obliged to fight for it in courts of law using Western, representational means—or as Paddy Roe and Frans Hoogland explain, "living country" had to be turned into "cultivated country." Meanwhile, they offered a very different engagement with country to participants in the *Lurujarri Heritage Trail*.

More than thirty years of welcoming visitors to the land, showing them around and answering their questions, evidence a remarkable history of *Goolarabooloo* entertaining their own means of introducing "living country" through lived experience. Such transecological means are as sustainable as scientific reports, not least because they motivated a large number of non-Indigenous people to support the opposition movement. This might even suggest that their ways are more sustainable, since reports such as the one by Bradshaw and Fry could be ignored by proponents of the LNG precinct, while the Trail and its protest(ers) could not.

Western law and science severely impact on accounts of ontologically different lifeworlds in that they demand that research(ers) whitewash some of their intrinsic qualities. In my case, instead of opening up the transecological space I was to report on, "white magic" asked me to render some of its components invisible—to insist that the translation of the intrinsic qualities and heritage values of "living country" into Western scientific discourse turn "living country" into "cultivated country," as if the facts thus obtained were always "out there." The main thing that is cultivated when doing so, however, is a "lasting" frontier between transecology and representationism, and between Settler-colonialism and its "otherwise."

7.3 TRANSLATION AS THE "LAST FRONTIER"

The above gives a good explanation of why the challenges that resource exploration initiatives pose for Indigenous people are significantly more complex and cannot be answered with scientific assessments alone. They demand that ontologically different world(view)s be taken seriously and performed in their own terms, which in the case of "living country" is through lived experience. Such world(view)s only begin where Western science reaches its limits. They cannot be represented or translated. However, this must not be misinterpreted as meaning that they are not of equal significance.

The "white magic" of modernity, of Western science and law, turns out to be magical only in so far as it renders other world(view)s invisible. Another magic is therefore needed: projects that focus on the entanglements of people and place that emphasize their intra-actions and diverse world-making activities. Viewed through such a transecological lens, the successful opposition movement makes a case for the recognition of "living country" as that which facilitated better-informed economic and political decision-making that was ultimately beneficial for all parties involved: without the protests of the *Goolarabooloo* and their collaborators, construction of the LNG facility at Walmadany/James Price Point would have commenced in 2009, and not only would the environment and what it holds have been destroyed as a consequence, but given the significant drop in oil and gas prices that followed, Woodside might well be bankrupt by now.

In this sense, the protest was successful on multiple fronts. However, the disturbances the conflict caused within and beyond the wider Broome community cannot be overestimated. Mr Roe died less than a year after Woodside announced its retreat, and with his death the *Goolarabooloo* lost their immediate leader, which in turn had an undeniable impact on not only the loss of their Native Title claim which followed in 2017, but also their general standing among the Indigenous communities in the area. The wider Broome community was similarly affected through the ways in which friendships were sought, maintained or ended because of the conflict. Martin Pritchard provided me with the opportunity to reflect on this myself during the interview I conducted with him:

> *Martin Pritchard:* It was extremely stressful, a lot of it (. . .) Because not only did we have the community but there was also obviously outside organizations that were campaigning on it. We had to merge the community together as a collective, and then also merge outside campaigns so they weren't creating any difficulties for the community side of things (. . .)
>
> *Carsten Wergin:* I spoke with other people who were heavily involved and also mentioned the stress factor. They were like, "Every morning, 3:00 or 4:00 in the morning you get a phone call, a text message, or a call and then you're in your car, you are picking people up and you are driving. Every day you are thinking, 'Okay, am I supposed to go out there?' You always have your swag prepared, your stuff prepared." Somebody told me about lavender bags that she put in her pants when she would do a lock-on. So it didn't smell when she would pee in her pants. [Then] the creativity involved... coming up with all these ideas of how to block the road and how to do a lock-on and this kind of thing.
>
> *MP:* That's right, it's been amazing. That's why I'm thinking they picked on the wrong community, really. There's a lot of strong opinion. It's characters here. Like everything in the Kimberley. The feeling I got when I first got here was everything was so big! The landscape, the issues, the threats of mining, etcetera, also the people. One of the first people in the first few months I met was [Mr Roe]. I was told not to meet him alone, "because he's dangerous" (. . .) Someone who worked for a particular organization had a run-in with him (. . .) Then you meet people like Louise Middleton [co-founder of the No Gas Campaign, together with Ebony L.] and other people in the community

> that are pretty solid and strong and opinionated, and won't back down. (Transcript 130513_1_MP)

The previous chapters have introduced many of these strong characters, who continue to fight for living heritage until this day. Four years after Woodside abandoned its project proposal for Walmadany/James Price Point, the *Goolarabooloo* lost their Native Title case and were therefore not recognized by the court as the custodians of the land in question. Industrialization is again looming, as the new Native Title holders have expressed a general openness to proposals. This might also be tied to or preceded by a discontinuation of the *Lurujarri Heritage Trail*. But rather than speculating about what might never be, I conclude with radical hope and the invocation of an Australian Dreaming after the resources boom.

Chapter 8

All Heritage Is Collaborative

> The turtles shed a tear when they are caught. And [Aboriginal people] speak to them, "You got the seagrass to eat, and we got this to eat" (. . .) When I was standing there someone said, "You know it looks cruel but this is how we have been doing it."
>
> (2015 Trail Participant, Transcript 150823_anonymous)

The above recollection by a 2015 Trail participant highlights how the *Lurujarri Heritage Trail* experience encourages a radical but equally realistic understanding of sustainability. The killing of the turtle is justified by a shared engagement with the land and sea that recognizes needs and sacrifices made for each other and on equitable terms. The *Lurujarri Heritage Trail,* viewed this way, becomes a pathway toward a possible common multispecies future, motivated by an Australian Dreaming that can inspire well beyond Australian shores.

For me, it inspired the notion of transecology that became so helpful when I needed to protect my ethnographic data from representationisms that discarded the relevance of experience, emphasizing the object under study over the process of objectification. In this way, transecology shows the resourceness of a place and how it unfolds in intra-action, highlighting the endurance of human and more-than-human collaborators of which such a place is constituted in the face of Settler-colonialist over-determinacy.

The fight for Walmadany/James Price Point provides one of many examples of places where people not only question promises of economic opportunity but openly express their preference for preserving what they have. Meanwhile, the policies and practices of late liberalism turn further away from the needs and demands of such people, whether they are environmentalists or Indigenous custodians of land and sea. Because of this, the pursuit of opportunities continues to not only justify many short-sighted decisions but also cause devastation among people and places for the benefit of the Settler-colonialist world(view).

The resources industry and WA state politics tried to infiltrate a local community on various levels with such promises, for the supposed benefit of the general population—for proper schooling and medical attention, to rescue the economy that was termed "in danger" and unable to be saved by tourism alone. They did not shy away from downplaying scientific results about the environmental impact of the industrialization proposal; these were ignored or eliminated, as such disturbances could negatively affect the "fair go" that was promoted in 2017 by then Prime Minister Turnbull.

With this in mind, "opportunity dreamers" found themselves backed in pursuing the industrialization of Walmadany/James Price Point by the egalitarian ideal of Multicultural Australia, with the "fair go" as its unifying component. Then protest actions began, drawing attention to the transecological values of the "otherwise": of the endurance of "living country" and living heritage, burial grounds, spirits, monsoonal vine thickets, dinosaur footprints, turtles, bilbies and humpback whales that are already in place.

Multifaceted protests highlighted the environmental, social, and cultural impact development would have on the wider region, and how this would affect its "nowhere else but here" qualities. Love and life—a communal spirit of "We love Broome," the *liyan* that people draw from "living country," and the values of its living heritage—were mobilized to demonstrate that there are transecological alternatives. Those who walked on "living country," who engaged and connected rather than talked or flew over it, produced more adequate accounts of its values. These might have been citizen scientists, casual tourists and travelers, or participants in the *Lurujarri Heritage Trail*.

The living heritage that this constitutes is very real because participating in the Trail means performing its reality. The Trail is a world-making activity that does not sugarcoat the fragility of its lifeworld, or the vulnerability of its diverse components, but opens up a transecological perspective on both. Based on these, further collaborations came into being that are done rather than spoken of.

The Australian Dreaming is a call for further such ventures into the "otherwise" that can facilitate decision-making that caters for both development and preservation. Conflicts between preservation and development will remain. Their numbers are even likely to increase. But there is hope, since the transecological approach can also be applied elsewhere in order to flesh out the radical potential of conflict situations that are toxic to the late-liberal mainstream and enable, to paraphrase Basil Sansom (2001), "irruptions of the otherwise."

Walmadany/James Price Point could not be overridden by the opportunity dreaming of an LNG facility that was to bury it under 15 meters of concrete. Its preservation therefore marks a crucial first: a general turn within contemporary Australian society toward values that are intrinsic to the land. This turn is strongly linked to Aboriginal knowledge, which has nurtured a more inclusive understanding of human and more-than-human intra-action, as Aboriginal people say, "since day one." It is also linked to hidden qualities of Indigenous tourism experiences such as the *Lurujarri Heritage Trail*. Their emphasis on (Indigenous) particularities offers people an alternative means of confronting issues of heritage preservation and industrial development, while their non-hierarchical regimes of value allow collaborators to depart from conventional calls for the recognition of a Multicultural Australia and instead position Indigenous and non-Indigenous collaborators in a transecological project that takes the entanglement of multiple world(view)s seriously.

Such transecological awareness is both locally situated and globally inclusive in terms of reclaiming the authority to define what is "good" beyond the generalizing technicalities of Western law, the over-determinism of representational modes of science, or the opportunity dreamings of late-liberal market economics. The Australian Dreaming instead proposes the pursuit of a transecological ethics, fostering onto-epistemic partnerships, putting processuality before progress and more-than-human concerns before human aspirations. Awareness of the many misrepresentations of Indigenous and non-Indigenous world(view)s that do not conform to the late-liberal mainstream plays a significant role in making this "otherwise" possible. In this way, the Australian Dreaming addresses the greatest contemporary challenge: the project of decolonization.

The successful protest for Walmadany/James Price Point highlights many such misrepresentations, but also suggests ways to overcome them. Among others, the *Lurujarri Heritage Trail*, Walmadany/James Price Point, and the *Goolarabooloo* are examples of how to think "Indigeneity" in unconventional, collaborative terms, while their emphasis on the entanglement of people and land sets the stage for a widespread and diversified awareness for the environmental, social, and cultural impacts of the proposed industrial development.

Such awareness brings the global resources boom to a turning point, as more people express their resentment of it, fostered by new collaborations—assemblages of diverse actors that articulate powerful responses to doubtful promises of unlimited wealth and progress. Development proposals can be conceived differently in such assemblages:

as transecological projects that tie the resourceness of a place to its lived experience. Indigenous tourism can make a substantial contribution to this by offering people possibilities for intra-action with this transecological "otherwise," with "living country" and its living heritage, that is not dead, but hiding, waiting to come back to the fore. In the end, such recognition might comply with the Dreaming that the late Paddy Roe sent out and walked on: a Dreaming that will help Australia evolve into a new global phase, which not only draws on a unique transecological history but also makes a crucial contribution to it.

Appendix

Tourism and Mining Tables

TOURISM

Table A1. Tourism Tables (National).

Total tourism employees[i]	*End of 2016: 580,200 (5% of the Australian workforce)*			
Visitor numbers The Source provides further information on each month, and a possibility to select the country of origin[ii]	Year ending December 2016: 8.270.500 Year ending December 2017: 8.810.800[iii] 914,500 visitor arrivals during February 2018 (up 16.1 % from previous year)			
Purpose of visit[iv] (In thousand people)	*YEAR*	*2016*	*2017*	*Increase by*
	Holiday	3.821	3.888	2%
	Visiting friends and relatives	1.902	2.228	17%
	Business	814	928	14%
	Education	495	553	12%
	Employment	279	254	–9%
	Other purpose	314	267	15%
Earnings from international visitors	$40.6 billion until December 2016[v] $41.3 billion until December 2017 (up 6%)[vi]			
GDP share [vii] (In Billion US Dollar)	Y2012: 1566.53; Y2013: 1515.77; Y2014: 1454.96; Y2015: 1232.92; Y2016: 1264.94			

(*continued*)

Table A1. (Continued)

Total tourism employees[i]	*End of 2016: 580,200 (5% of the Australian workforce)*		
Country of residence[viii] (In thousand)	*YEAR*	*2016*	*2017*
	China	1.116	1.251
	NZ	1.221	1.231
	US	668	729
	Japan	382	398
	Singapore	383	382
	Malaysia	347	353
	India	244	283
	S. Korea	262	277
	Germany	196	200
	Indonesia	156	173
	Canada	142	158
	France	117	122
	Scandinavia	102	105

Destinations of international visitors (Total number of visitors by state is greater than total number of visitors for Australia, as some visitors visited multiple states)[ix]	*YEAR*	*2016*	*2017*	*Increase by*
	NSW	3.872	4.158	7%
	VIC	2.674	2.891	8%
	QLD	2.574	2.684	4%
	WA	954	946	–1%
	SA	432	462	7%
	NT	287	293	2%
	TAS	236	279	18%
	ACT	208	243	17%

[i] https://www.theguardian.com/environment/2018/feb/08/climate-change-tourism-is-australias-least-prepared-industry-report-says#img-1.
[ii] http://www.tourism.australia.com/en/markets-and-research/tourism-statistics/international-visitor-arrivals.html.
[iii] http://www.tourism.australia.com/en/markets-and-research/tourism-statistics/international-visitor-arrivals.html.
[iv] https://www.tra.gov.au/tra/ivs/ivs_dec_2017.html.
[v] https://www.theguardian.com/environment/2018/feb/08/climate-change-tourism-is-australias-least-prepared-industry-report-says#img-1.
[vi] http://www.tourism.australia.com/en/markets-and-research/tourism-statistics/international-visitor-arrivals.html.
[vii] https://www.statista.com/statistics/263573/gross-domestic-product-gdp-of-australia/.
[viii] Tourism Western Australia. 2018. International Markets Summary: Year Ending December 2017. https://www.tourism.wa.gov.au/Publications%20Library/Research%20and%20reports/2017/International%20markets%20summary%20YE%20Dec%2017.pdf.
[ix] https://www.tra.gov.au/tra/ivs/ivs_dec_2017.html.
Source: Carsten Wergin.

Table A2. Tourism Tables for Western Australia.

	Before 2017	*From 2017 to 2018*
Visitors	1.4 million interstate (-8.2%); 8.1 million intrastate (-3.9%)[i] 953,800 international visitors (+9,1%) YE March 2017[ii]	"In 2017, 19.3 million intrastate daytrips were taken within the State and 10.6 million overnight (domestic and international) visitors came to or travelled within WA. Together, these visitors spent $9.6 billion in the State. While WA saw a 1.5% increase in total visitors for holiday purposes."[iii]
Earnings (as of March 2017) Estimates must be considered with confidence intervals: Spend ± 5.0%, Visitors ± 9.3%, Nights ± 9.2%	$1,526 million from interstate visitors $4,026 million from intrastate visitors[iv] $ 2,400 million from international visitors[v]	(in $million) YE December 2016; 2,417 YE December 2017: 2,297 (down 5%)[vi] Total visitors (international, national, domestic) $9,576 million in spend (down 4.7%)[vii]
Country of residence	Interstate visitors[viii]: New South Wales 32%; Victoria 31%, Queensland 18%, South Australia 12%, Tasmania 3%, Northern Territory 3%, ACT 2% International visitors[ix]: 953,800 (12.3% of international visitors to Australia) United Kingdom 16%, Malaysia 12%, Singapore 11%, USA 8%, New Zealand 8%, China 5%, Germany 4%, Indonesia 3%, Japan 3%, Hong Kong 2%, Other Asia 20%, Other Countries 5%	Until December 2017[x]: China 58,600 (+14.2%), Germany 32,900 (-15.6%), Hong Kong 24,000 (+5.7%), India 27,600 (+21.1%), Indonesia 30,200 (+7.1%), Japan 27,400 (+9.2%) Korea 10,300 (-36.0%), Malaysia 111,400 (+2.1%) NZ 68,900 (-11.3%), Singapore 102,000 (-2.1%), Switzerland 11,300 (-11%), UK 146,400 (-3.0%) USA 78,700 (-0.6%)

[i] Tourism Western Australia. 2017. Domestic Visitation: Fast Facts Year Ending March 2017.
[ii] Tourism Western Australia. 2017. International Visitation: Fast Facts YE March 2017.
[iii] Tourism Western Australia. 2018. Visitation to Western Australia: Overview Year ending December 2017. (*Source*: Tourism Research Australia (TRA) National Visitor Survey (NVS)), https://www.tourism.wa.gov.au/Publications%20Library/Research%20and%20reports/2017/Visitation%20to%20WA_OVERVIEW_YE%20Dec%2017%20v3.pdf.
[iv] Tourism Western Australia. 2017. Domestic Visitation: Fast Facts Year Ending March 2017.
[v] Tourism Western Australia. 2017. International Visitation: Fast Facts YE March 2017.
[vi] https://www.tra.gov.au/tra/ivs/ivs_dec_2017.html.
[vii] Tourism Western Australia. 2018. Visitation to Western Australia: Overview Year ending December 2017. (*Source*: Tourism Research Australia (TRA) National Visitor Survey (NVS)), https://www.tourism.wa.gov.au/Publications%20Library/Research%20and%20reports/2017/Visitation%20to%20WA_OVERVIEW_YE%20Dec%2017%20v3.pdf.
[viii] Tourism Western Australia. 2017. Domestic Visitation: Fast Facts Year Ending March 2017.
[ix] Tourism Western Australia. 2017. International Visitation: Fast Facts YE March 2017.
[x] Tourism Western Australia. 2018. International Markets Summary: Year Ending December 2017.
Source: Carsten Wergin.

Table A3. Tourism Tables for Northwest Australia (Kimberley and Pilbara).

Visitors	*Domestic visitors:*
Base for % is from total estimated international visitor numbers to Australia's Northwest. Estimates need to be considered with confidence Intervals: Spend ± 5.0%, Visitors ± 9.3%, Nights ± 9.2%	YE 2016 1,236,500 (intra 991,800; inter 244,700) YE 2017 1,172,500 (intra 973,300; inter 199,200) International visitors: YE 2016: 72,400 YE 2017: 61,600[i]
Earnings	Total overnight visitors northwest spent ($million): YE December 2016: 1,090 YE December 2017: 696 Domestic YE16 1.018, YE17 629 International YE16: 72; YE17: 67[ii]
Country of residence	YE 2016: New Zealand 12,900 (17%); USA 9,000 (12%); UK 8,500 (11%); Germany 8,200 (11%); France (5%)[iii] YE 2017: USA 11,200 (16%); NZ 10,500 (15%); UK 8,700 (12%); Germany 8,200 (12%); France 3,900 (6%)[iv]
Areas visited	Estimated intrastate visitors: Shire of East Pilbara 265,800 (27%), Shire of Ashburton 257,400 (27%), City of Karratha 157,300 (16%), Shire of Broome 144,900 (15%), Town of Port Hedland 132,300 (14%) Estimated Interstate Visitors: Shire of Broome 89,300 (39%), Shire of Wyndham-East Kimberley 57,400 (25%), Shire of Ashburton 42,200 (19%), Shire of Derby-West Kimberley 34,100 (15%), City of Karratha 33,600 (15%) Estimated International Visitors: Shire of Broome 34,600 (49%), Shire of Ashburton 25,100 (36%), City of Karratha 15,200 (22%), Town of Port Hedland 14,700 (21%), Shire of Wyndham-East Kimberley 13,200 (19%)[v]

[i] Tourism Western Australia. 2018. Visitation to Western Australia: Overview Year ending December 2017. (*Source*: Tourism Research Australia's (TRA) National Visitor Survey (NVS)), https://www.tourism.wa.gov.au/Publications%20Library/Research%20and%20reports/2017/Visitation%20to%20WA_OVERVIEW_YE%20Dec%2017%20v3.pdf.

[ii] Tourism Western Australia. 2018. Visitation to Western Australia: Overview Year ending December 2017. (*Source*: Tourism Research Australia's (TRA) National Visitor Survey (NVS)), https://www.tourism.wa.gov.au/Publications%20Library/Research%20and%20reports/2017/Visitation%20to%20WA_OVERVIEW_YE%20Dec%2017%20v3.pdf.

[iii] Tourism Western Australia. 2015. Australia´s North West: Overnight visitor Fact Sheet 2014/2015.

[iv] Tourism Western Australia. 2018. Australia´s North West 2017 Factsheet. https://www.tourism.wa.gov.au/Publications%20Library/Research%20and%20reports/2017/Tourism%20Regions%202017/ANW_2017%20Factsheet.pdf.

[v] Tourism Western Australia. 2018. Australia´s North West 2017 Factsheet. https://www.tourism.wa.gov.au/Publications%20Library/Research%20and%20reports/2017/Tourism%20Regions%202017/ANW_2017%20Factsheet.pdf.

Source: Carsten Wergin.

Table A4. Tourism Tables for Broome/Dampier Peninsula.

Visitors	*Broome Visitors:*
For detailed information such as on age, accommodation choice, and purpose of visit see Source[i]	2007: 146,600 Holiday or Leisure; 37,900 Business 2013: 113,500 Holiday and Leisure, 33,200 Business[ii] Broome estimated visitors YE Dec15/16/17: Intrastate 144,900 (54%) Interstate 89,300 (33%) International 34,600 (13%) Total 268,800 (100%) Broome estimated visitor nights YE Dec15/16/17: Intrastate 689,300 Interstate 683,900 International 331,300 Total 1,704,500 Broome Average Length of Stay by estimated nights (YE Dec15/16/17): Intrastate: 4.8; Interstate 7.7; International 9.6; Total 6.3 Peak of leisure traveling: 65% in September 2012, with 35% business. Off Peak leisure traveling: 41% in April 2012, while 59% business[iii] "Since the GFC, campers decreased due to a fall in the number of European travellers and backpackers. In 2012, visitors generally spent less time in Broome and attend fewer activities compared to previous years because they were more price-conscious."[iv]
Earnings	"The expenditure associated with visitors to Broome during 2010-11 was in excess of $160 million. The main source of this expenditure was from domestic overnight visitors. Visitor spending from tourism is injected directly into the local economy and benefits numerous small businesses across the area."[v] Broome Tourism Expenditure (2010-11) in average daily expenditure and total visitor expenditure: Daytrip visitors: $95; $M 3.3 Domestic overnight visitors: $168; $M 154.6 International Visitors $51; $M 9.2[vi]

(*continued*)

Table A4. (Continued)

Visitors	*Broome Visitors:*
	Annual hotel occupancy and revenue: 2007: ca. 76% occupancy, ca. 140$ revenue 2010: ca. 54%, ca. 97$ 2012: ca. 57%, ca. 110$[vii]

[i] Tourism Western Australia. 2018. Shire of Broome: Visitor Factsheet: Three Year Average 2015/2016/2017.
[ii] Haeberlin consulting. 2014. Broome Tourism Strategy: Final September 2014.
[iii] Haeberlin consulting. 2014. Broome Tourism Strategy: Final September 2014.
[iv] Haeberlin consulting. 2014. Broome Tourism Strategy: Final September 2014.
[v] AEC Group. 2012. Broome Economic Profile REVISED: Final Draft Report. http://www.broome.wa.gov.au/files/assets/public/planning/lps/economic-profile.pdf.
[vi] AEC Group. 2012. Broome Economic Profile REVISED: Final Draft Report. http://www.broome.wa.gov.au/files/assets/public/planning/lps/economic-profile.pdf.
[vii] Haeberlin consulting. 2014. Broome Tourism Strategy: Final September 2014.
Source: Carsten Wergin.

MINING

Table A5. Mining Tables (National).

General information	"In the year to June 2008-09, the mining industry recorded a \$47.2 billion increase in total income, a rise of 34.7%. Despite an increase in total expenses of \$25.5b (32.6%), operating profit before tax rose \$17.7b (38.6%)."[i] This was followed by an "industry value added decrease of 7.4% (\$9.4b) to \$118.3b in 2014-15, with the largest decrease in the Metal ore mining subdivision (12.9% or \$9b), related to falling commodity prices."[ii] "The Mining division Export Price Index fell 21% between 2013-14 and 2014-15, driven by falling prices for high quality metallurgical coal, thermal coal and iron ore [. . .]. Employment and wage growth also decreased over this period."[iii] "In 2016, Australia's mineral exports (excluding petroleum products) amounted to over \$151 billion, which was almost 58% of all export merchandise and 46% of all exported goods and services. In the 2016–17 financial year, mining accounted for 7.4% of gross domestic product."[iv] "Australia had a large number of mines producing gold (130), black coal (93), copper (34), iron ore (29) and silver (22). And few operating mines producing nickel (13), zinc (13), lead (13), zircon (8), ilmenite (7), rutile (7), bauxite (6), uranium (3), manganese ore (1) and diamond (1)."[v] "Using 2016 rates of production at operating mines, Australia's Measured and Indicated Resources of bauxite, black coal, copper, lead, manganese ore, silver, uranium and zinc could potentially last approx. 40 years. Operating gold mines have enough Measured and Indicated Resources to last an average of 22 years. Operating lode-gold deposits, which provide most of the gold production, have an average resource life of 8 years."[vi]
Employment	2014–2015: Total Mining 173,388 (down 6.9%; in 2013–2014: 186,194)[vii] "Structural changes as a result of the contraction of the Mining industry have continued to dampen employment demand. Employment fell 6.9% in 2014-15 (12,806 people) [. . .]. The biggest decrease was seen in Exploration and other mining support services (8,396 people or 19.3%)."[viii] Coal mining total 39,128 (down 9.8 % from year before) Metal ore mining total 65,035 (up 0.4 %) Non-metallic mining and quarrying 12,046 (down 1.1%)

(*continued*)

Table A5. (Continued)

	Exploration and mining support services 35,179 (down 19.3%) Oil and gas extraction 22,001 (down 1.2%) "In 2015-16, The Mineral Commodities sector employed around 228,000. In 2014-15, the sector paid $26 billion in wages and salaries."[ix]
Australia's Share of World EDR For EDR trends on specific Minerals see Source[x] For Numbers on EDR per Commodity see Source[xi]	"Australia's EDR of gold, iron ore, lead, rutile, zircon, nickel, uranium and zinc are the world's largest, while antimony, bauxite, black coal, recoverable brown coal, cobalt, copper, diamond, ilmenite, lithium, manganese, niobium, silver, tantalum, tungsten and vanadium all rank in the top six worldwide."[xii] "Australia has a large percentage of some of the world's economic demonstrated resources (EDR's)."[xiii] "Australia's EDR of gold, iron ore, lead, nickel, rutile, tantalum, uranium, zinc and zircon are the world's largest." "EDR's are proven or probable resource reserves. As at December 2008, Australia had 38% of the EDR (including resources which are inaccessible for mining because of environmental restrictions, government policies or military lands) for Nickel, 38% for Uranium, 33% for Lead, 28% for Zinc and 25% for Brown Coal. Other significant amounts of the world's EDR (including inaccessible resources) reserves within Australia include 20% of Silver, 16% of Industrial Diamonds, 15% of Iron Ore and 13% of Gold."[xv] "As of 31 December 2016, Australia's EDR increased for black coal, cobalt, gold, ilmenite, lithium, molybdenum, potash, rutile, tantalum, tin, vanadium and zircon. The strongest gains in EDR were mineral sands (11%), lithium (70%), potash (73%), cobalt (13%), tantalum (9%) and tin (12%) as improved markets for these commodities, particularly for those associated with battery technology, has stimulated exploration and resource delineation. The most significant fall in EDR was diamond (-45%) as the Argyle mine in WA winds down."[xvi]

[i] Mining statistics I news What's the latest in Mining statistics http://www.abs.gov.au/ausstats/abs@.nsf/Latest-products/8418.0Main%20Features52009%20to%202010?opendocument&tabname=Summary&prodno=8418.0&issue=2009%20to%202010&num=&view=.

[ii] Mining Operations, Australia, key figures 2014–2015 Latest issue Released at 11:30 a.m. (Canberra time) 27/05/2016 Final http://www.abs.gov.au/ausstats/abs@.nsf/0/D96FCC4AEEA50923CA2568A90013940B?Opendocument.

[iii] Mining Operations, Australia, key figures 2014–2015 Latest issue Released at 11:30 AM (Canberra time) 27/05/2016 Final http://www.abs.gov.au/ausstats/abs@.nsf/0/D96FCC4AEEA50923CA2568A90013940B?Opendocument.

[iv] Australian Government: Geoscience Australia. Australia's Identified Mineral Resources 2017. http://www.ga.gov.au/__data/assets/pdf_file/0005/58874/Australias-Identified-Mineral-Resources-2017.pdf.

[v] Australian Government: Geoscience Australia. Australia's Identified Mineral Resources 2017. http://www.ga.gov.au/__data/assets/pdf_file/0005/58874/Australias-Identified-Mineral-Resources-2017.pdf.

[vi] Australian Government: Geoscience Australia. Australia's Identified Mineral Resources 2017. http://www.ga.gov.au/__data/assets/pdf_file/0005/58874/Australias-Identified-Mineral-Resources-2017.pdf.

[vii] Mining Operations, Australia, key figures 2014–20015 Latest issue Released at 11:30 a.m. (Canberra time) 27/05/2016 Final http://www.abs.gov.au/ausstats/abs@.nsf/0/D96FCC4AEEA50923CA2568A90013940B?Opendocument.

Table A5. (Continued)

[viii] Mining Operations, Australia, key figures 2014–2015 Latest issue Released at 11:30 a.m. (Canberra time) 27/05/2016 Final http://www.abs.gov.au/ausstats/abs@.nsf/0/D96FCC4AEEA50923CA2568A90013940B?Opendocument.

[ix] Australian Government, Department of Industry, Innovation and Science. Australian Mineral Commodities. https://industry.gov.au/resource/Mining/AustralianMineralCommodities/Pages/default.aspx; Australian Government, Department of Industry, Innovation and Science. December 2016. Australia's Major Export Commodities: Overview. https://industry.gov.au/resource/Mining/AustralianMineralCommodities/Documents/Australias-major-export-commodities-overview.pdf.

[x] Geoscience Australia. Trends in Australian's Economic Demonstrated Resources of Major Mineral Commodities http://www.ga.gov.au/data-pubs/data-and-publications-search/publications/aimr/trends; http://www.ga.gov.au/data-pubs/data-and-publications-search/publications/aimr/execuite-summary.

[xi] Australian Government: Geoscience Australia. Australia's Identified Mineral Resources 2017. http://www.ga.gov.au/__data/assets/pdf_file/0005/58874/Australias-Identified-Mineral-Resources-2017.pdf.

[xii] http://www.ga.gov.au/data-pubs/data-and-publications-search/publications/aimr/execuite-summary.

[xiii] Mining statistics I news What's the latest in Mining statistics http://www.abs.gov.au/ausstats/abs@.nsf/Latestproducts/8418.0Main%20Features52009%20to%202010?opendocument&tabname=Summary&prodno=8418.0&issue=2009%20to%202010&num=&view=.

[xiv] Australian Government: Geoscience Australia. Australia's Identified Mineral Resources 2017. http://www.ga.gov.au/__data/assets/pdf_file/0005/58874/Australias-Identified-Mineral-Resources-2017.pdf.

[xv] Mining statistics I news What's the latest in Mining statistics http://www.abs.gov.au/ausstats/abs@.nsf/Latestproducts/8418.0Main%20Features52009%20to%202010?opendocument&tabname=Summary&prodno=8418.0&issue=2009%20to%202010&num=&view=.

[xvi] Australian Government: Geoscience Australia. Australia's Identified Mineral Resources 2017. http://www.ga.gov.au/__data/assets/pdf_file/0005/58874/Australias-Identified-Mineral-Resources-2017.pdf.

Source: Carsten Wergin.

Table A6. Mining Tables for Western Australia.

General	"The value of WA's mineral and petroleum industry in 2015–16 was $87.9 billion, down 12 per cent from 2014–15. Compared with the value of the industry in 2005–06 at $43 billion, it remains more than twice what it was 10 years ago."[i] "The mineral sector accounted for almost 80 per cent, or $69.5 billion, of WA's resources industry value in 2015–16."[ii] In 2017, WA's mineral and petroleum industry reported sales of $108.8 billion. The 16 percent increase in year-on-year sales value was led by improving volumes and relatively steady iron ore prices, a strong growth in LNG volumes and the continued strength of the gold sector.
Employment * This figure comprises only operations subject to State petroleum legislation, and excludes LNG operations and land-based service operations.	"WA's mining industry directly employed an average of 102,258 people during 2015–16. This figure includes people employed in mineral exploration, construction of mine site infrastructure, mineral processing, mine site surveying, transport and catering – essentially those people operating on site."[iii] "Mining-related employment declined for the second consecutive year after peaking at 107,871 people in 2013–14."[iv] "WA's petroleum industry employed an average of 1380 people in 2015–16, down 33 per cent from an average of 2058 in 2014–15."
Petrol Sector See Source for more numbers on quantity and value[vi]	"The value of WA's petroleum sector, comprising crude oil, condensate, LNG, natural gas and LPG (butane and propane), was affected by declining crude oil prices, which averaged US$45.05 per barrel, down from US$76.70 in 2014–15. As a result, the value of the petroleum sector decreased almost 24 per cent to $18.4 billion in 2015–16."[vii] The petroleum sector accounts for 20 per cent of the total value of mineral and petroleum sales in WA. This makes it the second most valuable sector after iron ore. The sector was valued at $21.7 billion in 2017. "The volume of LNG produced in WA has increased 38 per cent in 2017 reaching 32.7 million tones. LNG is WA's most valuable petroleum product, accounting for 14 per cent of the State's total mineral and petroleum sales in 2017. The value of LNG sales value rose from $10.6 billion in 2016 to $14.9 billion in 2017."[viii]

(*continued*)

Table A6. (Continued)

Sales of Value See Source for numbers on quantity and value on more of the minerals[ix]	"Iron ore was WA's highest value commodity, accounting for more than $48 billion (71 per cent) of total mineral sales (2015/16). WA remained the world's largest exporter of iron ore, with sales volumes reaching 756 million tones, up from 719 million tones in 2014–15."[x] Mineral commodities dominate the industry, accounting for 80 per cent, or $87 billion, of the total sales value. Iron ore: 72 per cent of mineral sales and 58 per cent of overall sales of $62.9 billion in 2017, an increase of 14 per cent. Gold: 210 tonnes (6.76 million ounces) in 2017, a 6.7 per cent increase on 2016 volumes. Alumina and bauxite: 7% of mineral value. Sales volumes increased by close to 5%. Major West Australian Commodities: Petroleum $21.5b; Gold $11b; Alumins $5.8b; Nickel $2.2b; Iron Ore $63b[xi] Exploration WA (Mineral): WA $ 1028m (66%) of $1565m for Australian Total Exploration WA (Petroleum): WA $652m (47%) of $1376m for Australian Total[xii] Crude oil and Condensate: Australian total 105.m Barrels, WA share 69%, 72 Barrels Natural Gas Australian Total 104b cubic meters, WA share 48% 51b CbM[xiii]
*Exploration expenditure** *Exploration expenditure and drilling activity is an indicator of the health of the minerals and petroleum industry.[xiv] See source for comparison between states.[xv]	On a state by state basis, the 2012 calendar year recorded increases in exploration in WA (up 13% on 2011 to $2052.6 million) and Tasmania (up 5% on 2011 to $40.7 million). "The value of projects under construction or in the committed stage of development was an estimated $54 billion, down from $96 billion in September 2017. The value of planned or possible projects remained steady, from $49 billion to $50 billion over the same period."[xvi] Australia's mineral exploration expenditure was $1.75 billion in 2017, up from $1.42 billion in 2016. Western Australia contributed over $1.1 billion, with the gold and iron ore sectors attracting the largest share.

(continued)

Table A6. (Continued)

Gains and costs	"As of September 2016, WA had an estimated $97 billion worth of resource projects under construction or in the committed stage of development. A further $46 billion was identified as planned or possible projects."[xvii] In 2017, Private New Capital Investment: Australian Total $114b, WA share $29b (25%) Mining Investment: Australian total $39b, WA share $22b (57%)[xviii] In 2017, Mineral and Energy Exports: Australian Total $209b, WA share $113b (54%) In 2017, Economic Impact: Australia GDP 2016-17 $1693b WA GSP 2015-16 $233b; Mining Component of WA GSP 31% In 2017, Merchandise Exports: Australian Total $301b, WA share $121b (42%) WA exports of minerals and energy as % share of all State Merchandise exports: 90%[xx] "More than $20 billion was invested in WA's mining industry in 2017, representing just under 55 per cent of national expenditure. Nationally, mining investment has fallen over 60 per cent in four years. The decline has been slightly less severe for WA, with mining investment falling 56 per cent over the same period."[xxi] "As of March 2018, the value of projects under construction or in the committed stage of development was an estimated $54 billion, down from $96 billion in September 2017. The value of planned or possible projects remained steady, from $49 billion to $50 billion over the same period."[xxii]

[i] Government of Western Australia. 2016. Department of Mines and Petroleum. Statistics digest 2015-16. http://www.dmp.wa.gov.au/Documents/About-Us-Careers/Stats_Digest_2015-16.pdf.

[ii] Government of Western Australia. 2016. Department of Mines and Petroleum. Statistics digest 2015-16. http://www.dmp.wa.gov.au/Documents/About-Us-Careers/Stats_Digest_2015-16.pdf.

[iii] Government of Western Australia. 2016. Department of Mines and Petroleum. Statistics digest 2015-16. http://www.dmp.wa.gov.au/Documents/About-Us-Careers/Stats_Digest_2015-16.pdf.

[iv] Government of Western Australia. 2016. Department of Mines and Petroleum. Statistics digest 2015-16. http://www.dmp.wa.gov.au/Documents/About-Us-Careers/Stats_Digest_2015-16.pdf.

[v] Government of Western Australia. 2016. Department of Mines and Petroleum. Statistics digest 2015-16. http://www.dmp.wa.gov.au/Documents/About-Us-Careers/Stats_Digest_2015-16.pdf.

Table A6. (Continued)

[vi] Government of Western Australia, Department of Mines, Industry Regulation and Safety. Latest Statistics Releases. Review 2017. http://www.dmp.wa.gov.au/About-Us-Careers/Latest-Statistics-Release-4081.aspx.

[vii] Government of Western Australia. 2016. Department of Mines and Petroleum. Statistics digest 2015-16. http://www.dmp.wa.gov.au/Documents/About-Us-Careers/Stats_Digest_2015-16.pdf.

[viii] Government of Western Australia, Department of Mines, Industry Regulation and Safety. Latest Statistics Releases. Review 2017, http://www.dmp.wa.gov.au/About-Us-Careers/Latest-Statistics-Release-4081.aspx.

[ix] Government of Western Australia, Department of Mines, Industry Regulation and Safety. Latest Statistics Releases. Review 2017. http://www.dmp.wa.gov.au/About-Us-Careers/Latest-Statistics-Release-4081.aspx.

[x] Government of Western Australia. 2016. Department of Mines and Petroleum. Statistics digest 2015-16. http://www.dmp.wa.gov.au/Documents/About-Us-Careers/Stats_Digest_2015-16.pdf.

[xi] Government of Western Australia. Department of Mines, Industry Regulation and Safety. Quick resource facts. http://www.dmp.wa.gov.au/About-Us-Careers/Quick-resource-facts-3961.aspx.

[xii] Government of Western Australia. Department of Mines, Industry Regulation and Safety. Quick resource facts. http://www.dmp.wa.gov.au/About-Us-Careers/Quick-resource-facts-3961.aspx.

[xiii] Government of Western Australia. Department of Mines, Industry Regulation and Safety. Quick resource facts. http://www.dmp.wa.gov.au/About-Us-Careers/Quick-resource-facts-3961.aspx.

[xiv] Government of Western Australia. Department of Mines, Industry Regulation and Safety. http://www.dmp.wa.gov.au/About-Us-Careers/Latest-Resources-Investment-4083.aspx.

[xv] Australian's Mineral Resource Assessment 2012. Mineral exploration. http://www.ga.gov.au/data-pubs/data-and-publications-search/publications/australian-minerals-resource-assessment/mineral-exploration.

[xvi] Government of Western Australia. Department of Mines, Industry Regulation and Safety. http://www.dmp.wa.gov.au/About-Us-Careers/Latest-Resources-Investment-4083.aspx.

[xvii] Government of Western Australia. 2016. Department of Mines and Petroleum. Statistics digest 2015-16. http://www.dmp.wa.gov.au/Documents/About-Us-Careers/Stats_Digest_2015-16.pdf.

[xviii] Government of Western Australia. Department of Mines, Industry Regulation and Safety. Quick resource facts. http://www.dmp.wa.gov.au/About-Us-Careers/Quick-resource-facts-3961.aspx.

[xix] Government of Western Australia. Department of Mines, Industry Regulation and Safety. Quick resource facts. http://www.dmp.wa.gov.au/About-Us-Careers/Quick-resource-facts-3961.aspx.

[xx] Government of Western Australia. Department of Mines, Industry Regulation and Safety. Quick resource facts. http://www.dmp.wa.gov.au/About-Us-Careers/Quick-resource-facts-3961.aspx.

[xxi] Government of Western Australia. Department of Mines, Industry Regulation and Safety. http://www.dmp.wa.gov.au/About-Us-Careers/Latest-Resources-Investment-4083.aspx.

[xxii] Government of Western Australia. Department of Mines, Industry Regulation and Safety. http://www.dmp.wa.gov.au/About-Us-Careers/Latest-Resources-Investment-4083.aspx.

Source: Carsten Wergin.

Table A7. Mining Tables for Northwest Australia (Kimberley).

General information	"The Kimberley is identified with significant resource potential. Minerals, offshore gas and potentially onshore gas are to impact on future development. Established natural resource industries include diamonds, gold, iron ore and nickel.[i] Meanwhile, the $356 million total share of WA production from the Kimberley represented only 0.34% of the state total in 2016-17."[ii] Kimberley production by commodity share 2016-17: Diamonds: 75% Gold, Silver, Gemstones: 18% Nickel, Copper, Cobalt: 5% Construction Materials: 2% Crude Oil: 0%[iii]
Employment	Employment in the Mining Industry in the Kimberley in 2011: 860[iv] Share of Mining among all Industries of Employment: 2011: 5,8%; 2006: 4,3%[v]
Value of minerals and petroleum	"In 2012-13, total resources production in the Kimberley was valued at $1.1 billion. Iron ore accounted for $535 million (48 per cent). Diamonds and crude oil make up 32 per cent of resources production, valued at $360 million. The remaining 20 per cent valued at $217 million, is comprised of nickel, copper, cobalt, gold, silver and construction materials."[vi] By region 2017: Wyndham-East Kimberley 268.511.784 Hall Creek 80.295.681 Derby-West Kimberley 12.763.079 Broome 3.625.339 By Commodity for whole Kimberley Region in 2017: Diamonds 267.062.677 Gold and Silver 79.409.596 Petroleum 12.569.716 Construction Materials and Gem and Semi Precious Stones 5.154.940 Nickel, Copper and Cobalt 875.941 Limes and Limestone Dolomite and Dimension Stone 123.014 Total: 365.195.884[vii] The value of resource production by shire in 2016-17 was $356 million (The Pilbara was $63 billion): • Wyndham-East Kimberley: $270 million • Halls Creek: $81 million • Broome: $4 million • Derby-West Kimberley: $1 million[viii] The value of resource production by commodity in 2016-17 • Diamonds: $268 million • Gold, silver and gemstones: $65 million • Nickel, copper, cobalt: $17 million • Construction materials: $5 million • Crude oil: $1m[ix]

(*continued*)

Table A7. (Continued)

[i] Government of Western Australia. Department of Regional Development. Kimberley: a region in profile 2014. http://kdc.wa.gov.au/wp-content/uploads/2016/08/Kimberley-A-Region-in-Profile-2014.pdf.

[ii] Kimberley Development Commission. 2018. Minerals and Energy. http://kdc.wa.gov.au/economic-activity/minerals-energy/.

[iii] Kimberley Development Commission. 2018. Minerals and Energy. http://kdc.wa.gov.au/economic-activity/minerals-energy/.

[iv] Kimberley Development Commission . 2016. Employment. http://kdc.wa.gov.au/economic-activity/employment/.

[v] Government of Western Australia. Department of Regional Development. Kimberley: a region in profile 2014. http://kdc.wa.gov.au/wp-content/uploads/2016/08/Kimberley-A-Region-in-Profile-2014.pdf.

[vi] Government of Western Australia. Department of Regional Development. Kimberley: a region in profile 2014. http://kdc.wa.gov.au/wp-content/uploads/2016/08/Kimberley-A-Region-in-Profile-2014.pdf.

[vii] Government of Western Australia. Department of Mines, Industry Regulation and Safety. 2018. A snapshot of the Kimberley Resources Industry. http://www.dmp.wa.gov.au/Documents/Community-Education/A_snapshot_of_the_Kimberley_Resources_Industry_Fact_Sheet.pdf.

[viii] Government of Western Australia. Department of Mines, Industry Regulation and Safety. 2018. A snapshot of the Kimberley Resources Industry. http://www.dmp.wa.gov.au/Documents/Community-Education/A_snapshot_of_the_Kimberley_Resources_Industry_Fact_Sheet.pdf.

[ix] Government of Western Australia. Department of Mines, Industry Regulation and Safety. 2018. A snapshot of the Kimberley Resources Industry. http://www.dmp.wa.gov.au/Documents/Community-Education/A_snapshot_of_the_Kimberley_Resources_Industry_Fact_Sheet.pdf.

Source: Carsten Wergin.

Table A8. Mining Tables for Broome/Dampier Peninsula.

General	"The economy of Broome is reliant on tourism, with retail trade, transport, accommodation and food services contributing roughly 30% to its GRP and employment in the area (2016). Broome's pearling industry has been severely affected by the global financial crisis and subsequent European debt crisis. Its cattle industry decreased as a result of the banning of live cattle exports from Australia to Indonesia and subsequently, after the ban was lifted, lower demand from Indonesia for Australian cattle."[i]
Value of employment	"In 2016, there was no employment in the mining industry within the Broome Local Government Area (LGA), resulting in a gross value add per employee of $0; GRP 0%, Employment 0%, Business Mix 1%."[ii]
Projects close to Broome For the Thunderbird mineral sand project, close to Broome, using Broome Port, see Source[iii]	Existing Projects: Ungani oilfield (halfway between Broome and Derby, south east of the Great Northern Highway), operated by Buru Energy. Commercial oil production started in July 2015 and is currently being produced and transported out of Wyndham Port. New Projects: "Thunderbird mineral sands project (halfway between Broome and Derby, north west of the Great Northern Highway). In 2018, Sheffield Resources is going through environmental approvals and Native Title processes and plans to start works in the same year, with first production planned for some time during the second half of 2019 with a mine life of more than 40 years." "Mineral sands are used for paints and plastics, ceramics, sunscreen, toothpaste, artificial joints and welding rods."[iv] "The Thunderbird deposit is the first major mineral sands deposit to be discovered in the Canning Basin and one of the largest mineral sands deposits to be discovered in the last 30 years."[v] "The project requires the clearing of up to 1961 hectares of native vegetation within a 5648 ha development envelope. The company would produce 12 to 18 million tones per annum of zircon and ilmenite product. [. . .] Existing facilities would be utilized for additional shipping through the Port of Broome."[vi]

(continued)

Table A8. (Continued)

Council regulations *Data reviewed 2017; for more on policies, see Source[vii]	"The Shire of Broome does not approve or regulate mining, petroleum or geothermal energy resource projects, but may be requested to provide comment or engage with proponents or the State in this regard."[viii]* Shire objectives: "To support the expansion of industry, and sustainable industry, that helps deliver economic prosperity to its residents, whilst ensuring every effort is undertaken to make sure that the initial assessment as well as compliance and monitoring of any Mining, Petroleum and Geothermal activities carried out are at a level that is compatible with the amenity of the natural and built environment and that of the Shire's residents."[ix] "To promote that any activities associated with mining or the extraction of petroleum and geothermal energy resources are to be thoroughly assessed, monitored and managed effectively by the appropriate agencies and to ensure local economic, environmental and social values are enhanced and developed."[x]

[i] AEC Group. 2012. Broome Economic Profile: Shire of Broome – Final draft report. http://www.broome.wa.gov.au/files/assets/public/planning/lps/economic-profile.pdf.

[ii] AEC Group. 2012. Broome Economic Profile: Shire of Broome – Final draft report. http://www.broome.wa.gov.au/files/assets/public/planning/lps/economic-profile.pdf.

[iii] Government of Western Australia. Department of Mines and Petroleum. Thunderbird mineral sand project.

[iv] Government of Western Australia. Department of Mines, Industry Regulation and Safety. 2018. A snapshot of the Kimberley Resources Industry, http://www.dmp.wa.gov.au/Documents/Community-Education/A_snapshot_of_the_Kimberley_Resources_Industry_Fact_Sheet.pdf.

[v] Sheffield Resources Limited. "Thunderbird." http://www.sheffieldresources.com.au/irm/content/thunderbird.aspx?RID=393.

[vi] Cordingley, Glenn. 2017. "Green light for mine proposal." https://thewest.com.au/news/regional/green-light-for-mine-proposal-ng-b88623099z.

[vii] Office of the Chief Executive Officer Organisational. 2015 (2017). Mining, petroleum and geothermal energy resource extraction. http://www.broome.wa.gov.au/files/assets/public/council/policy/policy-2018/125-mining-petroleum-and-geothermal-energy-resource-extraction.pdf?BestBetMatch=mining|b71d2268-146a-48ca-9783-01d4e0a71bae|970b6047-7dac-4fd2-a973-a10801713593.

[viii] Office of the Chief Executive Officer Organisational. 2015 (2017). Mining, petroleum and geothermal energy resource extraction, http://www.broome.wa.gov.au/files/assets/public/council/policy/policy-2018/125-mining-petroleum-and-geothermal-energy-resource-extraction.pdf?BestBetMatch=mining|b71d2268-146a-48ca-9783-01d4e0a71bae|970b6047-7dac-4fd2-a973-a10801713593.

[ix] Office of the Chief Executive Officer Organisational. 2015 (2017). Mining, petroleum and geothermal energy resource extraction, http://www.broome.wa.gov.au/files/assets/public/council/policy/policy-2018/125-mining-petroleum-and-geothermal-energy-resource-extraction.pdf?BestBetMatch=mining|b71d2268-146a-48ca-9783-01d4e0a71bae|970b6047-7dac-4fd2-a973-a10801713593.

[x] Office of the Chief Executive Officer Organisational. 2015 (2017). Mining, petroleum and geothermal energy resource extraction, http://www.broome.wa.gov.au/files/assets/public/council/policy/policy-2018/125-mining-petroleum-and-geothermal-energy-resource-extraction.pdf?BestBetMatch=mining|b71d2268-146a-48ca-9783-01d4e0a71bae|970b6047-7dac-4fd2-a973-a10801713593.

Source: Carsten Wergin.

Bibliography

Abascal, Trinidad, Martin Fluker, and Min Jiang. 2015. "Domestic Demand for Indigenous Tourism in Australia: Understanding Motivations, Barriers, and Implications for Future Development." *Journal of Heritage Tourism* 10 (1): 1–20. doi:10.1080/1743873x.2014.921183.

ABC News. 2014. "EPA Reassessing Controversial James Price Point Site for Gas Development." *ABC News*, November 26, 2014. Accessed November 11, 2016. http://www.abc.net.au/news/2014-11-27/epa-assessing-james-price-point-site/5920816.

ABS Australian Bureau of Statistics, Government Australia. 2017. "Western Australia (STE): Economy and Industry." Accessed June 25, 2018. http://stat.abs.gov.au/itt/r.jsp?RegionSummary&region=5&dataset=ABS_REGIONAL_ASGS2016&geoconcept=ASGS_2016&measure=MEASURE&datasetASGS=ABS_REGIONAL_ASGS2016&datasetLGA=ABS_REGIONAL_LGA2017&regionLGA=LGA_2017&regionASGS=ASGS_2016.

ABS Government Australia. 2000. "Special Article – The Australian Mining Industry: From Settlement to 2000." Accessed June 25, 2018. http://www.abs.gov.au/ausstats/abs%40.nsf/94713ad445ff1425ca25682000192af2/93136e734ff62aa2ca2569de00271b10!OpenDocument.

ABS Government Australia. 2009. "Mining Statistics News: Release of 2008-9 Mining Industry Statistics." Accessed June 25, 2018. http://www.abs.gov.au/ausstats/abs@.nsf/Latestproducts/8418.0Main%20Features42009%20to%202010?opendocument&tabname=Summary&prodno=8418.0&issue=2009%20to%202010&num=&view=.

ABS Government Australia. 2010. "Mining Statistics News: What's the Latest in Mining Statistics." Accessed June 25, 2018. http://www.abs.gov.au/ausstats/abs@.nsf/Latestproducts/8418.0Main%20Features52009%20to%202010?opendocument&tabname=Summary&prodno=8418.0&issue=2009%20to%202010&num=&view=.

ABS Government Australia. 2016. "Mining Operations, Australia, Key Figures 2014-15." Accessed June 25, 2018. http://www.abs.gov.au/ausstats/abs@.nsf/0/D96FCC4AEEA50923CA2568A90013940B?Opendocument.

ABS Government Australia. 2017a. "Australian Industry, 2015–16 Key Industry Points." Accessed June 25, 2018. http://www.abs.gov.au/AUSSTATS/abs@.nsf/Lookup/8155.0Main+Features12015-16?OpenDocument.

ABS Government Australia. 2017b. "Australian Industry, 2016–17, 2016–17 Key Industry Points." Accessed June 25, 2018. http://www.abs.gov.au/AUSSTATS/abs@.nsf/mf/8155.0/.

ABS Government Australia. 2018. "Mineral and Petroleum Exploration, Australia, Sep 2018: Summary Commentary: Mineral Exploration (Other than for Petroleum)." Accessed June 25, 2018. http://www.abs.gov.au/ausstats/abs@.nsf/0/7008565B2DD42E23CA25718C00151590?Opendocument.

Adams, Kathleen M. 1997. "Ethnic Tourism and the Renegotiation of Tradition in Tana Toraja (Sulawesi, Indonesia)." *Ethnology* 36 (4): 309–20. doi:10.2307/3774040.

AEC Group. 2012. "Broome Economic Profile: Shire of Broome Final Draft Report, November 2012." Accessed May 3, 2018. http://www.broome.wa.gov.au/files/assets/public/planning/lps/economic-profile.pdf.

Akerman, Kim. 1981. *Horde Areas and Mythological Sites between James Price Point and Coconut Well on the West Coast of Dampierland*. Broome: Kimberley Land Council.

Altman, Jon. 1989. "Tourism Dilemmas for Aboriginal Australians." *Annals of Tourism Research* 16 (4): 456–76. doi:10.1016/0160-7383(89)90002-9.

Altman, Jon. 2004. "Economic Development and Indigenous Australia: Contestations over Property, Institutions and Ideology." *The Australian Journal of Agricultural and Resource Economics* 48 (3): 513–34. doi:10.1111/j.1467-8489.2004.00253.x.

Altman, Jon. 2007. "Alleviating Poverty in Remote Indigenous Australia: The Role of the Hybrid Economy." *Development Bulletin* 72. Accessed November 11, 2016. http://caepr.cass.anu.edu.au/research/publications/alleviating-poverty-remote-indigenous-australia-role-hybrid-economy.

Altman, Jon. 2010. "What Future for Remote Indigenous Australia? Economic Hybridity and the Neo-Liberal Turn." In *Culture Crisis – Anthropology and Politics in Aboriginal Australia*, edited by Jon Altman and Melinda Hinkson, 259–80. Sydney: University of New South Wales Press.

Altman, Jon, and Julie Finlayson. 1992. *Aborigines, Tourism and Sustainable Development*. Canberra: Centre for Aboriginal Economic Policy Research, Australian National University.

Appadurai, Arjun. 1986. "Introduction: Commodities and the Politics of Value." In *The Social Life of Things: Commodities in Cultural Perspective*, edited by Arjun Appadurai, 3–63. Cambridge: Cambridge University Press.

Appel, Hannah, Arthur Mason, and Michael Watts. 2015. *Subterranean Estates: Life Worlds of Oil and Gas*. Ithaca: Cornell University Press.

ATSIC and Office of National Tourism. 1997. *The National Aboriginal and Torres Strait Islander Tourism Strategy*. Canberra: ATSIC.

Attanapola, Chamila T., and Ragnhild Lund. 2013. "Contested Identities of Indigenous People: Indigenization or Integration of the Veddas in Sri Lanka." *Singapore Journal of Tropical Geography* 34 (2): 172–87. doi:10.1111/sjtg.12022.

Attwood, Bain. 1996. "The Past as Future: Aborigines, Australia and the (Dis) course of History." In *In the Age of Mabo: History, Aborigines and Australia*, edited by Bain Attwood, vii–xxxviii. Sydney: Allen & Unwin.

Australia Government. "Australian Atlas of Minerals Resources, Mines & Processing Centres." Accessed June 25, 2018. http://www.australianminesatlas.gov.au/history/index.html.

Barad, Karen. 2012. "Nature's Queer Performativity." *Kvinder, Køn & Forskning* 26 (1–2): 25–53. doi:10.7146/kkf.v0i1-2.28067.

Bastin, Rohan. 2003. "Surrender to the Market: Thoughts on Anthropology, the Body Shop and Intellectuals." *The Australian Journal of Anthropology* 14 (1): 19–38. doi:10.1111/j.1835-9310.2003.tb00218.x.

Bates, Daisy. (1947) 1985. *Native Tribes of Western Australia*, edited by Isobel White. Canberra: National Library of Australia.

Beckett, Jeremy. 2012. "Returned to Sender: Some Predicaments of Re-indigenisation." *Oceania* 82 (1): 104–12. doi:10.1002/j.1834-4461.2012.tb00122.x.

Behrends, Andrea. 2008. "Fighting for Oil When There is No Oil Yet: The Darfur-Chad Border." *Focaal* 52: 39–56. doi:10.3167/fcl.2008.520103.

Bennett, Jane. 2010. *Vibrant Matter: A Political Ecology of Things*. Durham: Duke University Press.

Benterrak, Krim, Stephen Muecke, and Paddy Roe. 1984. *Reading the Country: Introduction to Nomadology*. Perth: Fremantle Arts Centre Press.

Bessire, Lucas, and David Bond. 2014. "Ontological Anthropology and the Deferral of Critique." *American Ethnologist* 41 (3): 440–56. doi:10.1111/amet.12083.

Bird-David, Nurit. 1999. "Animism Revisited: Personhood, Environment and Relational Epistemology." *Current Anthropology* 40 (S1): 67–91. doi:10.1086/200061.

Blühdorn, Ingolfur, and Ian Welsh, eds. 2008. *The Politics of Unsustainability: Eco-Politics in the Post-Ecologist Era*. London: Routledge.

Blundell, Valda. 1993. "Aboriginal Empowerment and Souvenir Trade in Canada." *Annals of Tourism Research* 20 (1): 64–87. doi:10.1016/0160-7383(93)90112-g.

Bolaños, Omaira. 2011. "Redefining Identities, Redefining Landscapes: Indigenous Identity and Land Rights Struggles in the Brazilian Amazon." *Journal of Cultural Geography* 28 (1): 45–72. doi:10.1080/ 08873631.2011.548480.

Bradshaw, Elizabeth, and Rachel Fry. 1989. *Management Report for Lurujarri Heritage Trail*. Perth: Department of Aboriginal Sites, Western Australian Museum.

Brown, Denise. 1999. "Mayas and Tourists in the Maya World." *Human Organization* 58 (3): 295–304. doi:10.17730/humo.58.3.y21r558615941121.

Bruner, Edward M., and Barbara Kirshenblatt-Gimblett. 2005. "Maasai on the Lawn: Tourist Realism in East Africa." In *Culture on Tour: Ethnographies of Travel*, edited by Edward M. Bruner, 33–70. Chicago: Chicago University Press.

Burke, Paul. 2011. *Law's Anthropology: From Ethnography to Expert Testimony in Native Title*. Canberra: ANU Press. http://press.anu.edu.au?p=151301.

Burke, Peter. 2009. *Cultural Hybridity*. Cambridge: Polity Press.

Butler, Judith. 1997. *Excitable Speech: A Politics of the Performative*. New York and London: Routledge.

Butler, Richard, and Tom Hinch. 2007. "Indigenous Tourism: A Common Ground for Discussion." In *Tourism and Indigenous Peoples: Issues and Implications*, edited by Tom Hinch and Richard Butler, 1–15. Oxford: Butterworth-Heinemann.

Buutljens, Jeremy, Deborah Gale, and Nadine White. 2010. "Synergies between Australian Indigenous Tourism and Ecotourism: Possibilities and Problems for Future Development." *Journal of Sustainable Tourism* 18 (4): 497–513. doi:10.1080/09669581003653518.

Buzard, James. 1993. *The Beaten Track: European Tourism, Literature, and the Ways to Culture, 1800–1918*. Oxford: Oxford University Press.

Cappo Mike, Craig Syms, Marcus Stowar, Charlotte Johansson, and Tim Cooper. 2011. *Survey of Fish Habitat Associations in the Region Offshore from James Price Point Using Baited Remote Underwater Video Systems (BRUVS)*. Townsville: Australian Institute of Marine Science.

Carrington, Kerry, Russell Hogg, Alison McIntosh, and John Scott. 2012. "Crime Talk, FIFO Workers and Cultural Conflict on the Mining Boom Frontier." *Australian Humanities Review* 53. Accessed January 3, 2019. http://australianhumanitiesreview.org/2012/11/01/crime-talk-fifo-workers-and-cultural-conflict-on-the-mining-boom-frontier/.

Carse, Ashley. 2012. "Nature as Infrastructure: Making and Managing the Panama Canal Watershed." *Social Studies of Science* 42 (4): 539–63. doi:10.1177/0306312712440166.

Chabay, Ilan. 2010. "Voices at the Table: Participation, Ownership, and Collaboration in Social-Ecological issues." In *Overcoming the Challenges of 'Doing Participation' in Environment and Development*, edited by John Forrester and Åsa Gerger Swartling, 7–20. Stockholm: Stockholm Environment Institute Working Paper.

Clifford, James. 2013. *Returns: Becoming Indigenous in the Twenty-First Century*. Cambridge, MA: Harvard University Press.

Cohen, Erik. 1979. "Rethinking the Sociology of Tourism." *Annals of Tourism Research* 6 (1): 18–35. doi:10.1016/0160-7383(79)90092-6.

Coleman, Simon, and Mike Crang. 2002. *Tourism: Between Place and Performance*. Oxford: Berghahn Books.

Cordingley, Glenn. 2017a. "Green Light for Mine Proposal." *The West Australian*, October 8, 2017. Accessed June 25, 2018. https://thewest.com.au/news/regional/green-light-for-mine-proposal-ng-b88623099z.

Cordingley, Glenn. 2017b. "Mining Proposal Moves to Critical Stage." *The West Australian*, January 25, 2017. Accessed June 25, 2018. https://thewest.com.au/news/broome-advertiser/mining-proposal-moves-to-critical-stage-ng-b88359018z.

Coronil, Fernando. 1997. *The Magical State: Nature, Money, and Modernity in Venezuela*. Chicago: University of Chicago Press.

Cresswell, Tim, and Peter Merriman, eds. 2011. *Geographies of Mobilities: Practices, Spaces, Subjects*. Farnham: Ashgate Publishing.

Cronin, Darryl, and Carsten Wergin. 2012. "An Alternative Indigenous Research Strategy." *SPRC Newsletter* 110.

Crutzen, Paul J., and Eugene Stoermer. 2000. "The Anthropocene." *Global Change Newsletter* 41 (1): 17–18.

Davidov, Veronica. 2013. *Ecotourism and Cultural Production: An Anthropology of Indigenous Spaces in Ecuador*. New York: Palgrave MacMillan.

DeLanda, Manuel. 2000. *A Thousand Years of Non-Linear History*. Cambridge, MA: MIT Press.

Deleuze, Gilles, and Felix Guattari. 1988. *A Thousand Plateaus: Capitalism and Schizophrenia*. London: Athlone Press.

Department of Industry, Innovation and Science, Government Australia. 2016. "Australia's Major Export Commodities: Overview." Accessed June 25, 2018. https://archive.industry.gov.au/resource/Mining/AustralianMineralCommodities/Documents/Australias-major-export-commodities-overview.pdf.

Department of Industry, Innovation and Science, Government Australia. n.d. "Australian Mineral Commodities." Accessed June 25, 2018. https://archive.industry.gov.au/resource/Mining/AustralianMineralCommodities/Pages/default.aspx.

Department of Resources, Energy and Tourism. 2009. "National Long Term Tourism Strategy." Accessed June 25, 2018. https://www.austrade.gov.au/ArticleDocuments/5499/DRET%20Tourism%20Strategy.pdf.aspx.

Descola, Philippe, and Gísli Pálsson, eds. 1996. *Nature and Society: Anthropological Perspectives*. London: Routledge.

DMP Department of Mines, Industry Regulation and Safety, Government of Western Australia. n.d. "Latest Resource Investment." Accessed June 25, 2018. http://www.dmp.wa.gov.au/About-Us-Careers/Latest-Resources-Investment-4083.aspx.

DMP Department of Mines, Industry Regulation and Safety, Government of Western Australia. n.d. "Quick Resource Facts." Accessed June 25, 2018. http://www.dmp.wa.gov.au/About-Us-Careers/Quick-resource-facts-3961.aspx.

DMP Department of Mines, Industry Regulation and Safety, Government of Western Australia. n.d. "Thunderbird Mineral Sand Project." Accessed June 25, 2018. http://www.dmp.wa.gov.au/Documents/Investors/Thunderbird_Mineral_Sands_Project_INFO_Sheet.pdf.

DMP Department of Mines, Industry Regulation and Safety, Government of Western Australia. 2016. "Department of Mines and Petroleum. Statistics Digest 2015-16." Accessed June 25, 2018. http://www.dmp.wa.gov.au/Documents/About-Us-Careers/Stats_Digest_2015-16.pdf.

DMP Department of Mines, Industry Regulation and Safety, Government of Western Australia. 2017. "Economic Indicators Resources Data." Accessed June 25, 2018. http://www.dmp.wa.gov.au/About-Us-Careers/Latest-Statistics-Release-4081.aspx.

DMP Department of Mines, Industry Regulation and Safety, Government of Western Australia. 2018. "A Snapshot of the Kimberley Resources Industry." Accessed June 25, 2018. http://www.dmp.wa.gov.au/Documents/Community-Eduction/A_snapshot_of_the_Kimberley_Resources_Industry_Fact_Sheet.pdf.

Dombrowski, Kirk. 2007. "Subsistence Livelihood, Native Identity and Internal Differentiation in Southeast Alaska." *Anthropologica* 49 (2): 211–29. http://www.jstor.org/stable/25605359.

Dragovich, Deirdre. 1993. "Aboriginal Rock Art and Visitors to Mootwingee National Park." *Australian Aboriginal Studies* 2: 58–65.

Dumont, Louis. 1986. "Genesis, I – The Christian Beginnings: From the Outworldly Individual to the Individual-in-the-World." In *Essays on Individualism: Modern Ideology in Anthropological Perspective*, edited by Louis Dumont, 23–59. Chicago: University of Chicago Press.

Dürr, Eveline, and Arno Pascht, eds. 2017. *Environmental Transformations and Cultural Responses: Ontologies, Discourses and Practices*. New York: Palgrave Macmillan.

EPA. 2012. "EPA Approves First Derived Proposal for Browse LNG Development." Accessed November 11, 2016. http://www.epa.wa.gov.au/media-statements/epa-approves-first-derived-proposal-browse-lng-development.

Eriksen, Thomas Hylland. 2016. *Overheating: An Anthropology of Accelerated Change*. Chicago: Pluto Press.

Escobar, Arturo. 2001. "Culture Sits in Places: Reflections on Globalism and Subaltern Strategies of Localization." *Political Geography* 20 (2): 139–74. doi:10.1016/s0962-6298(00)00064-0.

Fisher, Roger, William Ury, and Bruce Patton. (1981) 2011. *Getting to Yes: Negotiating Agreement Without Giving In*, 3rd ed. New York: Penguin Books.

Foucault, Michel. 1978. *The History of Sexuality, Vol. 1: An Introduction*. New York: Vintage Books.

Gale, Fay, Roger A. Luebbers, Jacquie Gillen, and Kristin Scott. 1988. *Tourist Impact on Aboriginal Cultural Sites in the Flinders Ranges, South Australia*. Adelaide: Department of Geography, University of Adelaide.

Galliford, Mark. 2010. "Touring 'Country,' Sharing 'Home': Aboriginal Tourism, Australian Tourists and the Possibilities for Cultural Transversality." *Tourist Studies* 10 (3): 227–44. doi:10.1177/1468797611407759.

Geertz, Clifford. 1973. *The Interpretation of Cultures: Selected Essays*. New York: Basis Books.

Geoscience, Government Australia. 2012a. "Australia's Identified Mineral Resources 2012: Executive Summary." Accessed June 25, 2018. http://www.ga.gov.au/data-pubs/data-and-publications-search/publications/aimr/execuite-summary.

Geoscience, Government Australia. 2012b. "Australia's Identified Mineral Resources 2012: Trends in Australia's Economic Demonstrated Resources of Major Mineral Commodities." Accessed June 25, 2018. http://www.ga.gov.au/data-pubs/data-and-publications-search/publications/aimr/trends.

Geoscience, Government Australia. 2013a. "Australia's Mineral Resource Assessment 2013: Mineral Exploration." Accessed June 25, 2018. http://www.ga.gov.au/data-pubs/data-and-publications-search/publications/australian-minerals-resource-assessment/mineral-exploration.

Geoscience, Government Australia. 2013b. "Australia's Mineral Resource Assessment 2013: Introduction." Accessed June 25, 2018. http://www.ga.gov.au/

data-pubs/data-and-publications-search/publications/australian-minerals-resource-assessment/introduction.

Geoscience, Government Australia. 2017. "Australia's Identified Mineral Resources 2017." Accessed June 25, 2018. http://www.ga.gov.au/__data/assets/pdf_file/0005/58874/Australias-Identified-Mineral-Resources-2017.pdf.

Gibson, Chris, and John Connell. 2005. *Music and Tourism: On the Road Again*. Clevedon: Channel View Publications.

Gilberthorpe, Emma. 2007. "Fasu Solidarity: A Case Study of Kin Networks, Land Tenure, and Oil Extraction in Kutubu, Papua New Guinea." *American Anthropologist* 109 (1): 101–12. doi:10.1525/aa.2007.109.1.101.

Glowczewski, Barbara. 1998. "The Meaning of 'One' in Broome, Western Australia: From Yawuru Tribe to Rubibi Corporation." *Aboriginal History* 22: 203–22. doi:10.22459/ah.22.2011.12.

Goolarabooloo and Broome No Gas Community, and Malcolm Lindsay. 2011. "Evidence of the Greater Bilby, Macrotis Lagotis, at the Site of the Proposed James Price Point Browse LNG Precinct." Accessed December 8, 2015. https://cdn.wilderness.org.au/archive/pdf/james-price-point-bilby-report.

Goolarabooloo and Broome No Gas Community, Madeline Goddard, Charlotte Buckton, Lily Leahy, and Malcolm Lindsay. 2013. "A Community Survey of Humpback Whales, Megaptera Novaeangliae, Near the Site of the Proposed James Price Point Browse Liquefied Natural Gas Precinct." Accessed December 11, 2015. http://kimberleycommunitywhaleresearch.files.wordpress.com/2013/03/a-community-survey-of-humpback-whales-near-james-price-point-march-2013.pdf.

Guyer, Jane I. 2004. *Marginal Gains: Monetary Transactions in Atlantic Africa*. Chicago: University of Chicago Press.

Guyer, Jane I., and LaRay Denzer. 2009. "The Craving for Intelligibility: Speech and Silence on the Economy under Structural Adjustment and Military Rule in Nigeria." In *Economic Persuasions*, edited by Stephen Gudeman, 97–117. Oxford and New York: Berghahn Books.

Guyer, Jane I., and LaRay Denzer. 2013. "Prebendalism and the People: The Price of Petrol at the Pump." In *Democracy and Prebendalism in Nigeria: Critical Interpretations*, edited by Wale Adebanwi and Ebenezer Obadare, 53–78. New York: Palgrave MacMillan.

Haeberlin Consulting. 2014. "Broome Tourism Strategy: Final September 2014." Accessed January 8, 2019. https://www.tourism.wa.gov.au/Publications%20Library/Research%20and%20reports/Broome_Tourism_Strategy_Final_Report_Full.pdf.

Hall, Michael C., and Hazel Tucker. 2004. *Tourism and Postcolonialism: Contested Discourses, Identities and Representations*. London: Routledge.

Haraway, Donna. 2015. "Anthropocene, Capitalocene, Plantationocene, Chthulucene: Making Kin." *Environmental Humanities* 6 (1): 159–65. doi:10.1215/22011919-3615934.

Head, Brian W. 2008. "Wicked Problems in Public Policy." *Public Policy* 3 (2): 101–18.

Healy, Chris. 2008. *Forgetting Aborigines*. Sydney: University of New South Wales Press.

Helmreich, Stefan. 2011. "Nature/Culture/Seawater." *American Anthropologist* 113 (1): 132–44. doi:10.1111/j.1548-1433.2010.01311.x.

Hemming, Steve. 1994. "In the Tracks of Ngurunderi: The South Australian Museum's Ngurunderi Exhibition and Cultural Tourism." *Australian Aboriginal Studies* 2: 38–46.

Higgins-Desbiolles, Freya. 2003. "Reconciliation Tourism: Tourism Healing Divided Societies!" *Tourism Recreation Research* 28 (3): 35–44. doi:10.1080/02508281.2003.11081415.

Higgins-Desbiolles, Freya. 2009. "Indigenous Ecotourism's Role in Transforming Ecological Consciousness." *Journal of Ecotourism* 8 (2): 144–60. doi:10.1080/14724040802696031.

Hinch, Thomas D., Thomas A. Delamere, and D. G. Reid. 1993. "Native Festivals as Tourism Attractions: A Community Challenge." *Journal of Applied Recreation Research* 18 (2): 131–42.

Hohl, Andreas E., and Clem A. Tisdell. 1995. "Peripheral Tourism: Development and Management." *Annals of Tourism Research* 22 (3): 517–34. doi:10.1016/0160-7383(95)00005-Q.

Holbraad, Martin, and Morton A. Pedersen. 2017. *The Ontological Turn: An Anthropological Exposition*. Cambridge: Cambridge University Press.

Holbraad, Martin, Morton A. Pedersen, and Eduardo Viveiros de Castro. 2014. "The Politics of Ontology: Anthropological Positions." *Cultural Anthropology Online*. Accessed November 21, 2016. http://www.culanth.org/fieldsights/462-the-politics-of-ontology-anthropological-positions.

Holm, Poul, Michael Evan Goodsite, Sierd Cloetingh, Mauro Agnoletti, Bedrich Moldan, Daniel J. Lang, Rik Leemans, et al. 2013. "Collaboration between the Natural, Social and Human Sciences in Global Change Research." *Environmental Science & Policy* 28: 25–35. doi:10.1016/j.envsci.2012.11.010.

Honer, Anne. 2000. "Lebensweltanalyse in der Ethnografie." In *Qualitative Forschung: Ein Handbuch*, edited by Uwe Flick, Ernst von Kardorff, and Ines Steinke, 194–204. Hamburg: Rowohlt.

Hulme, Mike. 2011. "Meet the Humanities." *Nature Climate Change* 1 (4): 177–79. doi:10.1038/nclimate1150.

in't Veld, Roeland J. 2010. *Knowledge Democracy: Consequences for Science Politics, and Media*. Berlin: Springer.

Ingold, Timothy. 2009. "Against Space: Place, Movement, Knowledge." In *Boundless Worlds: An Anthropological Approach to Movement*, edited by Peter W. Kirby, 29–44. New York: Berghahn.

Jahn, Pilippa, and Brigid Drummond. n.d. *Butcher Joe Nangan: An Introduction*. Accessed January 9, 2019. http://desertriversea.com.au/download/Butcher%20Joe%20Nangan.%20An%20Introduction%20By%20Philippa%20Jahn%20and%20Brigid%20Drummond.pdf.

Jenner, Kenneth Curt Selby, Micheline-Nicole Jenner, and Kathryn A. McCabe. 2001. "Geographical and Temporal Movements of Humpback Whales in Western Australian Waters." *The APPEA Journal* 41 (1): 749–61. doi:10.1071/aj00044.

Jorgenson, Joseph G. 1990. *Oil Age Eskimos*. Berkeley: University of California Press.

Karpik, Lucien. 2010. *Valuing the Unique: The Economics of Singularities*. Princeton: Princeton University Press.

KDP Kimberley Development Commission, Government of Western Australia. 2014. "Kimberley: A Region in Profile 2014." Accessed June 25, 2018. http://kdc.wa.gov.au/wp-content/uploads/2016/08/Kimberley-A-Region-in-Profile-2014.pdf.

KDP Kimberley Development Commission, Government of Western Australia. 2016. "Employment." Accessed June 25, 2018. http://kdc.wa.gov.au/economic-activity/employment/.

Kimberley Development Commission. 2018. "Minerals and Energy." Accessed June 25, 2018. http://kdc.wa.gov.au/economic-activity/minerals-energy/.

King, Brian. 1994. "What is Ethnic Tourism? An Australian Perspective." *Tourism Management* 15 (3): 173–76. doi:10.1016/0261-5177(94)90101-5.

King, David A., and William P. Stewart. 1996. "Ecotourism and Commodification: Protecting People and Places." *Biodiversity & Conservation* 5 (3): 293–305. doi:10.1007/bf00051775.

Kirksey, S. Eben, Nicholas Shapiro, and Maria Brodine. 2013. "Hope in Blasted Landscapes." *Social Science Information* 52 (2): 228–56. doi:10.1177/0539018413479468.

Kolig, Erich. 2000. "Social Causality, Human Agency and Mythology: Some Thoughts on History-Consciousness and Mythical Sense among Australian Aborigines." *Anthropological Forum* 10 (1): 9–30. doi:10.1080/00664670050006730.

Kunkel, Benjamin. 2017. "The Capitalocene." *London Review of Books*, March 2, 2017. Accessed January 8, 2019. https://www.lrb.co.uk/v39/n05/benjamin-kunkel/the-capitalocene.

Lane, Ruth, and Gordon Waitt. 2001. "Authenticity in Tourism and Native Title: Place, Time and Spatial Politics in the East Kimberley." *Social & Cultural Geography* 2 (4): 381–405. doi:10.1080/14649360120092607.

Lash, Scott, and John Urry. 1994. *Economies of Signs and Space*. London: Sage.

Latour, Bruno. 2013. *An Inquiry into Modes of Existence: An Anthropology of the Moderns*. Cambridge, MA: Harvard University Press.

Latour, Bruno. 2017. *Facing Gaia: Eight Lectures on the New Climatic Regime*. Translated by Catherine Porter. Cambridge: Polity Press.

Lea, Tess. 2006. "Cars, Corporations, Ceremonies and Cash: Hidden Co-Dependencies in Australia's North." In *Moving Anthropology: Critical Indigenous Studies*, edited by Tess Lea, Eemma Kowal, and Gillian Cowlishaw, 37–54. Darwin: Charles Darwin University Press.

Lea, Tess. 2012. "Contemporary Anthropologies of Indigenous Australia." *Annual Review of Anthropology* 41: 187–202. doi:10.1146/annurev-anthro-092611-145734.

Lear, Johnathan. 2008. *Radical Hope: Ethics in the Face of Cultural Devastation*. Cambridge: Harvard University Press.

Loverseed, H. 1998. "Aboriginal Tourism in North America." *Travel & Tourism Analyst* 6: 42–61.

Lowenthal, David. 1998. *The Heritage Crusade and the Spoils of History*. Cambridge: Cambridge University Press.

Luhrmann, Baz. 2008. *Australia* (The Movie). Distributed by 20th Century Fox.

MacCannell, Dean. 1976. *The Tourist: A New Theory of the Leisure Class*. Berkeley: University of California Press.

Marcus, George E. 1995. "Ethnography in/of the World System: The Emergence of Multi-Sited Ethnography." *Annual Review of Anthropology* 24 (1): 95–117. doi:10.1146/annurev.anthro.24.1.95.

Marsh, Helene, Helen Penrose, Carol Eros, and Joanna Hughes. 2002. *Dugong: Status Reports and Action Plans for Countries and Territories*. Nairobi: United Nations Environment Program.

McGregor, William. 1988. "A Survey of the Languages of the Kimberley Region-Report from the KLRC." *Australian Aboriginal Studies* 2: 92–102.

Merlan, Francesca. 1996. "Formulations of Claim and Title: A Comparative Discussion." In *Heritage and Native Title: Anthropological and Legal Perspectives*, edited by Julie Finlayson and Ann Jackson-Nakano, 165–77. Canberra: Australian Institute of Aboriginal and Torres Strait Islander Studies.

Merlan, Francesca. 2005. "Indigenous Movements in Australia." *Annual Review of Anthropology* 34 (1): 473–94. doi:10.1146/annurev.anthro.34.081804.120643.

Mining Technology. 2018. "Thunderbird Mineral Sands Project, Dampier Peninsula." Accessed June 25, 2018. http://www.mining-technology.com/projects/thunderbird-mineral-sands-project-dampier-peninsula/.

Mitrofankin, Evgenii. 2006. "Oil Companies' Relations with Reindeer Herders and Hunters in Nogliskii District, Northeastern Sakhalin Island." *Sibirica* 5 (2): 139–52. doi:10.3167/136173606780490757.

Moore, Jason W. 2015. *Capitalism in the Web of Life: Ecology and the Accumulation of Capital*. New York: Verso.

Moreton-Robinson, Aileen M. 2003. "I Still Call Australia Home: Indigenous Belonging and Place in a White Postcolonising Society." In *Uprootings / Regroundings: Questions of Home and Migration*, edited by Sara Ahmed, Claudia Castañeda, Anne-Marie Fortier, and Mimi Sheller, 23–40. Oxford: Berg Publishers.

Morton, Timothy. 2013. *Hyperobjects: Philosophy and Ecology after the End of the World*. Minneapolis: University of Minnesota Press.

Muecke, Stephen. 1999. "Visiting Aboriginal Australia." *Postcolonial Studies* 2 (1): 49–54. doi:10.1080/13688799989887.

Muecke, Stephen. 2004. *Ancient and Modern: Time, Culture and Indigenous Philosophy*. Sydney: UNSW Press.

Muecke, Stephen. 2011. "Australian Indigenous Philosophy." *Comparative Literature and Culture* 13 (2): 1–7. http://dx.doi.org/10.7771/1481-4374.1741>.

Muecke, Stephen. 2012. "'I am What I am Attached To': On Bruno Latour's 'Inquiry into the Modes of Existence.'" *Los Angeles Review of Books*, December 28, 2012. Accessed November 30, 2016. https://lareviewofbooks.org/article/i-am-what-i-am-attached-to-on-bruno-latours-inquiry-into-the-modes-of-existence/#!.

Muecke, Stephen. 2016. *The Mother's Day Protest and Other Fictocritical Essays*. London: Rowman and Littlefield.

Muecke, Stephen. 2017. "Earthbound Law: The Force of an Indigenous Australian Institution." *Law & Critique* 28 (2): 135–43. doi:10.1007/s10978-017-9206-7.

Muecke, Stephen, and Ben Dibley. 2016. "Five Theses for Reinstituting Economics: Anthropological Lessons from Broome." In *Engaging Indigenous Economy: Debating Diverse Approaches*, edited by Will Sanders, 143–52. Canberra: ANU Press.

Muir, Kathie. 2012. "Politics, Protest and Performativity: The Broome Community's 'No Gas on the Kimberley Coast' Campaign." *Australian Humanities Review* 2012 (53). Accessed January 9, 2019. http://australianhumanitiesreview.org/2012/11/01/politics-protest-and-performativity-the-broome-communitys-no-gas-on-the-kimberley-coast-campaign/.

Munn, Nancy. 1970. "The Transformation of Subjects into Objects in Walbiri and Pitjantjatjara Myth." In *Australian Aboriginal Anthropology: Modern Studies in the Social Anthropology of the Australian Aborigines*, edited by Ronald M. Berndt, 141–63. Nedlands: University of Western Australia Press.

Mustoe, Simon, and Matt Edmunds. 2008. *Coastal and Marine Natural Values of the Kimberley*. Sydney: WWF-Australia. Accessed December 2, 2016. http://awsassets.wwf.org.au/downloads/wa019_coastal_and_marine_natural_values_kimberley_1jan08.pdf.

Myers, Fred R. 1987. "Representing Whom? Privilege, Position and Posturing: A Rejoinder." *Canberra Anthropology* 10 (1): 62–73.

Myers, Fred R. 1989. "Truth, Beauty and Pintupi Painting." *Visual Anthropology* 2: 163–95.

Myers, Fred R. 1991. "Representing Culture: The Production of Discourse(s) for Aboriginal Acrylic Paintings." *Cultural Anthropology* 6 (3): 26–62.

Myers, Fred R. 1994. "Culture-Making: The Performance of Aboriginality at the Asia Society Galleries." *American Ethnologist* 21: 679–99.

Myers, Fred R. 1998. "Uncertain Regard? Understanding an Exhibition of Aboriginal Art in France." *Ethnos* 63: 7–47.

Myers, Fred R. 2001. "Introduction: The Empire of Things." In *The Empire of Things*, edited by Fred R. Myers, 3–63. Santa Fe: School of American Research Press.

Neimanis, Astrida. 2015. "No Representation without Colonisation? (Or, Nature Represents Itself)." *Somatechnics* 5 (2): 135–53. doi:10.3366/soma.2015.0158.

Notzke, Claudia. 1999. "Indigenous Tourism Development in the Arctic." *Annals of Tourism Research* 26 (1): 55–76. doi:10.1016/s0160-7383(98)00047-4.

Office of the Chief Executive Officer Organisational. 2015/2017. "Mining, Petroleum and Geothermal Energy Resource Extraction." Accessed June 25, 2018. http://www.broome.wa.gov.au/files/assets/public/council/policy/policy-2018/125-mining-petroleum-and-geothermal-energy-resource-extraction.pdf?BestBetMatch=mining|b71d2268-146a-48ca-9783-01d4e0a71bae|970b6047-7dac-4fd2-a973-a10801713593|.

Ortiz, Fernando. 1995. *Cuban Counterpoint, Tobacco and Sugar*. Durham: Duke University Press.

Pahl-Wostl, Claudia, Carlo Giupponi, Keith Richards, Claudia Binder, Alex de Sherbinin, Detlef Sprinz, Theo Toonen, and Caroline van Bers. 2013. "Transition Towards a New Global Change Science: Requirements for Methodologies, Methods, Data and Knowledge." *Environmental Science & Policy* 28: 36–47. doi:10.1016/j.envsci.2012.11.009.

Papatheodorou, Andreas, Jaume Rosselló, and Honggen Xiao. 2010. "Global Economic Crisis: Consequences and Perspectives." *Journal of Travel Research* 49 (1): 39–45.

Pedersen, Howard, and Banjo Woorunmurra. 2000. *Jandamarra and the Bunuba Resistance*. Broome: Magabala Books Aboriginal Corporation.

Peterson, Nicolas. 2011. "Is the Aboriginal Landscape Sentient? Animism, the New Animism and the Warlpiri." *Oceania* 81 (2): 167–79. doi:10.1002/j.1834-4461.2011.tb00101.x.

Petri, Helmut. 1987. "Sogenannte 'Schamanistische' Erscheinungsformen im Heutigen Nordwest-Australien." *Anthropos* 82 (4/6): 614–21.

Petri, Helmut. 2011. *The Dying World in Northwest Australia*. Translated by Ian Campbell. Carlisle: Hesperian Press.

Plumwood, Val. 2009. "Nature in the Active Voice." *Australian Humanities Review* 46 (1): 113–29. doi:10.22459/ahr.46.2009.10.

Polanyi, Karl. 1944. *The Great Transformation*. New York: Farrar & Reinhart.

Povinelli, Elizabeth A. 1993. *Labor's Lot: The Power, History and Culture of Aboriginal Action*. Chicago: University of Chicago Press.

Povinelli, Elizabeth A. 1995. "Do Rocks Listen? The Cultural Politics of Apprehending Australian Aboriginal Labor." *American Anthropologist* 97 (3): 505–18. doi:10.1525/aa.1995.97.3.02a00090.

Povinelli, Elizabeth A. 1998. "The Cunning of Recognition: Real Being and Aboriginal Recognition in Settler Australia." *Australian Feminist Law Journal* 11 (1): 3–27. doi:10.1080/13200968.1998.11077279.

Povinelli, Elizabeth A. 2002. *The Cunning of Recognition: Indigenous Alterities and the Making of Australian Multiculturalism*. Durham: Duke University Press.

Povinelli, Elizabeth A. 2016. *Geontologies: A Requiem to Late Liberalism*. Durham: Duke University Press.

Pulitano, Elivira, ed. 2014. *Indigenous Rights in the Age of the UN Declaration*. Cambridge: Cambridge University Press.

Richardson, Tanya, and Gisa Weszkalnys. 2014. "Introduction: Resource Materialities." *Anthropological Quarterly* 87 (1): 5–30. doi:10.1353/anq.2014.0007.

Roe, Paddy, and Frans Hoogland. 1999. "Black and White, a Trail to Understanding." In *Listen to the People, Listen to the Land*, edited by James Sinatra and Phin Murphy, 11–30. Carlton: Melbourne University Press.

Rogers, Douglas. 2015. "Oil and Anthropology." *Annual Review of Anthropology* 44: 365–80. doi:10.1146/annurev-anthro-102214-014136.

Roon, Tatiana. 2006. "Globalization of Sakhalin's Oil Industry: Partnership or Conflict? A Reflection on the Etnologicheskaia Ekspertiza." *Sibirica* 5 (2): 95–114. doi:10.3167/136173606780490702.

Rose, Deborah Bird. 1996. *Nourishing Terrains: Australian Aboriginal Views of Landscape and Wilderness*. Canberra: Australian Heritage Commission.

Ross, Helen. 1991. "Controlling Access to Environment and Self: Aboriginal Perspectives on Tourism." *Australian Psychologist* 26 (3): 176–82. doi:10.1080/00050069108257245.

Ruhanen, Lisa, Michelle Whitford, and Char-lee McLennan. 2015a. "Indigenous Tourism in Australia: Time for a Reality Check." *Tourism Management* 48: 73–83. doi:10.1016/j.tourman.2014.10.017.

Ruhanen, Lisa, Michelle Whitford, and Char-lee McLennan. 2015b. "Exploring Chinese Visitor Demand for Australia's Indigenous Tourism Experiences." *Journal of Hospitality and Tourism Management* 24: 25–34. doi:10.1016/j.jhtm.2015.07.002.

Rumsey, Alan, and James Weiner. 2004. *Mining and Indigenous Lifeworlds in Australia and Papua New Guinea*. Wantage: Sean Kingston.

Ryan, Chris, and Jeremy Huyton. 2000. "Who is Interested in Aboriginal Tourism in the Northern Territory, Australia? A Cluster Analysis." *Journal of Sustainable Tourism* 8 (1): 53–88. doi:10.1080/09669580008667349.

Sahlins, Marshall. 1972. *Stone Age Economics*. Chicago: Aldine.

Sansom, Basil. 2001. "Irruptions of the Dreamings in Post-Colonial Australia." *Oceania* 72 (1): 1–32.

Sawyer, Suzana, and Edmund Terence Gomez. 2012. *The Politics of Resource Extraction: Indigenous Peoples, Multinational Corporations, and the State*. New York: Palgrave Macmillan.

Scherrer, Pascal, and Kim Doohan. 2014. "Taming Wicked Problems: Towards a Resolution of Tourism Access to Traditional Owner Lands in the Kimberley Region, Australia." *Journal of Sustainable Tourism* 22 (7): 1003–22. doi:10.1080/09669582.2013.847943.

Schmiechen, Joc. 2006. *Indigenous Tourism Research Agenda: Key Directions for the Future 2008–2015*. Gold Coast: Sustainable Tourism Cooperative Research Centre.

Schuler, Sigrid, Lucinda Aberdeen, and Pamela Dyer. 1999. "Sensitivity to Cultural Difference in Tourism Research: Contingency in Research Design." *Tourism Management* 20 (1): 59–70. doi:10.1016/s0261-5177(98)00093-4.

Scranton, Roy. 2013. "Learning How to Die in the Anthropocene." *New York Times*, November 10, 2013. Accessed December 2, 2016. http://opinionator.blogs.nytimes.com/2013/11/10/learning-how-to-die-in-the-anthropocene/?_php=true&_type=blogs&_r=0.

Sheffield Resources Limited. 2018. "Thunderbird." Accessed June 25, 2018. http://www.sheffieldresources.com.au/irm/content/thunderbird.aspx?RID=393.

Sheller, Mimi, and John Urry. 2006. "The New Mobilities Paradigm." *Environment & Planning A: Economy and Space* 38: 207–26.

Sheller, Mimi, and John Urry. 2016. "Mobilizing the New Mobilities Paradigm." *Applied Mobilities* 1: 10–25.

Short, Damien. 2003. "Reconciliation, Assimilation, and the Indigenous Peoples of Australia." *International Political Science Review* 24 (4): 491–513. doi:10.1177/01925121030244005ü.

Smith, Benjamin R. 2007. "Towards an Uncertain Community? The Social Effects of Native Title in Central Cape York Peninsula." In *The Social Effects of Native Title: Recognition, Translation, Coexistence*, edited by Benjamin R. Smith and Frances Morphy, 117–34. Canberra: ANU Press.

Sofield, Trevor H. B. 2002. "Australian Aboriginal Ecotourism in the Wet Tropics Rainforest of Queensland, Australia." *Mountain Research and Development* 22 (2): 118–22. doi:10.1659/0276-4741(2002)022[0118:aaeitw]2.0.co;2.

Stammler, Florian. 2005. *Reindeer Nomads Meet the Market: Culture, Property and Globalisation at the 'End of the Land.'* Münster: LIT Verlag.

Stammler, Florian, and Vladislav Peskov. 2008. "Building a 'Culture of Dialogue' among Stakeholders in North-West Russian Oil Extraction." *Europe-Asia Studies* 60 (5): 831–49. doi:10.1080/09668130802085182.

Stanner, William Edward Hanley. 1965. "Aboriginal Territorial Organization: Estate, Range, Domain and Regime." *Oceania* 36 (1): 1–26. doi:10.1002/j.1834-4461.1965.tb00275.x.

Stanner, William Edward Hanley. 1979. *White Man Got No Dreaming: Essays 1938–1973*. Canberra: ANU Press.

Statista. n.d. "Gross Domestic Product of Australia." Accessed May 3, 2018. https://www.statista.com/statistics/263573/gross-domestic-product-gdp-of-australia/.

Steinberg, Michael K. 1994. "Tourism Development and Indigenous People." *Focus on Geography* 44 (2): 17–20. doi:10.1111/j.1949-8535.1994.tb00080.x.

Stokes, Bronwyn, and William B. McGregor. 2003. "Classification and Subclassification of the Nyulnyulan Languages." In *The Non-Pama–Nyungan Languages of Northern Australia: Comparative Studies of the Continent's Most Linguistically Complex Region*, edited by Nicholas Evans, 29–74. Canberra: Pacific Linguistics.

Swain, Tony. 1993. *A Place for Strangers: Towards a History of Australian Aboriginal Being*. Cambridge: Cambridge University Press.

Swyngedouw, Erik. 2010. "Impossible Sustainability and the Postpolitical Condition." *Urban and Landscape Perspectives* 9 (2): 185–205.

Tàbara, J. David, and Ilan Chabay. 2013. "Coupling Human Information and Knowledge Systems with Social–Ecological Systems Change: Reframing Research, Education, and Policy for Sustainability." *Environmental Science & Policy* 28: 71–81. doi:10.1016/j.envsci.2012.11.005.

Taylor-Gooby, Peter. 2013. "Why Do People Stigmatise the Poor at a Time of Rapidly Increasing Inequality, and What can be Done about It?" *The Political Quarterly* 84 (1): 31–42. doi:10.1111/j.1467-923x.2013.02435.x.

The Australian Government. 2017. "Multicultural Australia: United, Strong, Successful." Accessed January 9, 2019. https://www.homeaffairs.gov.au/mca/Statements/english-multicultural-statement.pdf.

The Pigram Brothers. 1997. *Saltwater Country* (Music Record). Jigil Records.

The Wilderness Society, Environs Kimberley, Conservation Council of WA. 2012. "James Price Point Science Assessment Report: Updated Supplementary Submission on the WA Government's Browse LNG Strategic Assessment for a Proposed Gas Processing Hub at James Price Point, Kimberley WA." Submission to the WA EPA and Commonwealth Department of Environment (SEWPaC). Accessed December 2, 2016. https://www.wilderness.org.au/files/james-price-point-science-assessment-report.

Tindale, Norman B. 1974. *Aboriginal Tribes of Australia: Their Terrain, Environmental Controls, Distribution, Limits, and Proper Names*. Berkeley: University of California Press.

Todd, Zoe. 2016. "An Indigenous Feminist's Take on the Ontological Turn: 'Ontology' is Just another Word for Colonialism." *Journal of Historical Sociology* 29 (1): 4–22. doi:10.1111/johs.12124.

Tonkinson, Robert. 2007. "Aboriginal 'Difference' and 'Autonomy' Then and Now: Four Decades of Change in a Western Desert Society." *Anthropological*

Forum: A Journal of Social Anthropology and Comparative Sociology 17 (1): 41–60. doi:10.1080/00664670601168476.

Tourism, Government of Western Australia. 2015a. "Australia´s North West: Overnight Visitor Fact Sheet 2014/2015."

Tourism, Government of Western Australia. 2015b. "Kimberley Tourism Facts and Figures."

Tourism, Government of Western Australia. 2017a. "Domestic Visitation: Fast Facts Year Ending March 2017."

Tourism, Government of Western Australia. 2017b. "International Visitation: Fast Facts YE March 2017."

Tourism, Government of Western Australia. 2018a. "Understanding Repeat Visitation to Western Australia: Summary March 2018." Accessed May 3, 2018. https://www.tra.gov.au/ArticleDocuments/258/WA_Repeat_Visitation_March_2018_FINAL.PDF.aspx?Embed=Y.

Tourism, Government of Western Australia. 2018b. "Visitation to Western Australia: Overview: Year Ending December 2017." Accessed May 3, 2018. https://www.tourism.wa.gov.au/Publications%20Library/Research%20and%20reports/2017/Visitation%20to%20WA_OVERVIEW_YE%20Dec%2017%20v3.pdf.

Tourism, Government of Western Australia. 2018c. "Australia´s North West 2017 Factsheet." Accessed May 3, 2018. https://www.tourism.wa.gov.au/Publications%20Library/Research%20and%20reports/2017/Tourism%20Regions%202017/ANW_2017%20Factsheet.pdf.

Tourism, Government of Western Australia. 2018d. "International Markets Summary: Year Ending December 2017." Accessed May 3, 2018. https://www.tourism.wa.gov.au/Publications%20Library/Research%20and%20reports/2017/International%20markets%20summary%20YE%20Dec%2017.pdf.

Tourism, Government of Western Australia. 2018e. "Shire of Broome: Visitor Factsheet: Three Year Average – 2015/2016/2017." Accessed May 3, 2018. https://www.tourism.wa.gov.au/Publications%20Library/Research%20and%20reports/2017/LGAs%20and%20sub-regions/Broome_2017%20Factsheet.pdf.

Tourism Australia. "International Visitors Arrivals." Accessed May 3, 2018. http://www.tourism.australia.com/en/markets-and-research/tourism-statistics/international-visitor-arrivals.html.

Tourism Research Australia. 2011. *Snapshots 2011: Indigenous Tourism Visitors in Australia*. Canberra: Department of Resources, Energy and Tourism.

TRA Tourism Research, Government Australia. 2011. "The Economic Importance of Tourism in Australia's Regions." Accessed May 3, 2018. https://www.ecotourism.org.au/assets/Resources-Hub-Ecotourism-Research/gg.pdf.

TRA Tourism Research, Government Australia. 2017. "International Visitors in Australia." Accessed May 3, 2018. https://www.tra.gov.au/ArticleDocuments/242/IVS_one_pager_Dec2017_FINAL.pdf.aspx?Embed=Y.

TRA Tourism Research, Government Australia. 2018. "International Visitors in Australia: December 2017 Year Ending Results of the International Visitors Survey." Accessed May 3, 2018. https://www.tra.gov.au/tra/ivs/ivs_dec_2017.html.

Tsing, Anna L. 2015a. *The Mushroom at the End of the World: On the Possibility of Life in Capitalist Ruins.* Princeton: Princeton University Press.

United Nations. 2009. *State of the World's Indigenous Peoples.* New York: Department of Economic and Social Affairs New York, United Nations.

UNWTO (United Nations World Tourism Organization). 2009. "Ministers' Summit Focusing on Roadmap for Recovery: Sustainable Tourism in Challenging Times," July 9, 2009. Accessed December 5, 2012. http://www.hospitalitynet.org/news/4042418.html.

Van Dooren, Thom. 2014. *Flight Ways: Life and Loss at the Edge of Extinction.* New York: Columbia University Press.

Van Dooren, Thom, Eben Kirksey, and Ursula Münster. 2016. "Multispecies Studies Cultivating Arts of Attentiveness." *Environmental Humanities* 8 (1): 1–23.

Vermeersch, Laura, Dale Sanders, and Greg Willson. 2016. "Generation Y: Indigenous Tourism Interests and Environmental Values." *Journal of Ecotourism* 15 (2): 184–98. doi:10.1080/14724049.2016.1165233.

WA Government. 2011. "Browse LNG Precinct Strategic Assessment Report." Accessed December 15, 2015. http://www.dsd.wa.gov.au/state-development-projects/lng-precincts/browse-kimberley/browse-lng---environment.

Wall, Geoffrey. 1996. "Perspectives on Tourism in Selected Balinese Villages." *Annals of Tourism Research* 23 (1): 123–37. doi:10.1016/0160-7383(95)00056-9.

Watts, Michael. 2012. "A Tale of Two Gulfs: Life, Death, and Dispossession along Two Oil Frontiers." *American Quarterly* 64 (3): 437–67. doi:10.1353/aq.2012.0039.

Weaver, David B. 2016. "Beyond Convention: Reimagining Indigenous Tourism." *Tourism Geographies* 18 (1): 28–37. doi:10.1080/14616688.2015.1122077.

Wells, Patricia Atkinson, ed. 1996. *Keys to the Marketplace: Problems and Issues in Cultural and Heritage Tourism.* Middlesex: Hisarlik Press.

Welsch, Wolfgang. 1999. "Transculturality – the Puzzling Form of Cultures Today." In *Spaces of Culture: City, Nation, World,* edited by Mike Featherstone and Scott Lash, 194–213. London: Sage.

Wergin, Carsten. 2010a. "Sounding Traditions that Do Not Exist: The Case of Réunion Island Music." *Orientwissenschaftliche Hefte* 28: 105–20.

Wergin, Carsten. 2010b. *Kréol Blouz: Musikalische Inszenierungen von Identität und Kultur* (with CD). Cologne, Weimar, and Vienna: Böhlau.

Wergin, Carsten. 2012. "Trumping the Ethnic Card: How Tourism Entrepreneurs on Rodrigues Tackled the 2008 Financial Crisis." *Island Studies Journal* 7 (1): 119–34.

Wergin, Carsten. 2015. "Travelling the Mascarenes: Creoleness in Tourism Policies and Practices on La Réunion, Mauritius and Rodrigues." In *Archipelago Tourism: Policies and Practices,* edited by Godfrey Baldacchino, 227–40. Farnham: Ashgate.

Wergin, Carsten. 2016. "Dreamings Beyond 'Opportunity': The Collaborative Economics of an Aboriginal Heritage Trail." *Journal of Cultural Economy* 9 (5): 488–506. doi:10.1080/17530350.2016.1210532.

Wergin, Carsten. 2017a. "Reconstructing Biodiversity for Tourism Development: Ethnographic Accounts from a World Heritage Site in the Making." In *World Heritage Sites and Tourism: Global and Local Relations,* edited by Maria Gravari-Barbas, Laurent Bourdeau, and Mike Robinson, 105–16. New York: Routledge.

Wergin, Carsten. 2017b. "The 'White Magic' of Modernity: Retracing Indigenous Environmental Knowledge in Settler-Colonialist Australia." In *Environmental Transformations and Cultural Responses: Ontologies, Discourses and Practices in Oceania*, edited by Eveline Dürr and Arno Pascht, 157–85. New York: Palgrave Macmillan.

Wergin, Carsten. 2018a. "Collaborations of Biocultural Hope: Community Science Projects against Industrialisation in Northwest Australia." *Ethnos: Journal of Anthropology* 83 (3): 455–72.

Wergin, Carsten. 2018b. "Policy in the Anthropocene." *Glocalism: Journal of Culture, Politics and Innovation* 2018 (3): 1–16. doi:10.12893/gjcpi.2018.3.3.

Wergin, Carsten. 2018c. "Tourism Meets Environmental Activism: Ecologies of Protest at Manari Road, Western Australia (August 2012)." In *Nature and Environment in Australia*, edited by Beate Neumeier, Boris Braun, and Victoria Herche, 103–19. Trier: WVT.

Wergin, Carsten. 2019. "Ecologies of the Otherwise." In *Ecocritical Concerns and the Australian Continent*, edited by Beathe Neumeier and Helen Tiffin, 109–22. Lanham: Lexington Books.

Wergin, Carsten. 2021. "Responsibility = Ownership? An Ethnographic Moment in Native Title." In *Mabo's Cultural Legacy. History, Literature, Film and Cultural Practice in Contemporary Australia*, edited by Geoff Rodoreda and Eva Bischoff, 47–58. London: Anthem Press.

Wergin, Carsten, and Corinna Erckenbrecht. 2018. *Der Ruf des Schneckenhorns: Hermann Klaatsch (1863 – 1916). Ein Heidelberger Wissenschaftler in Nordwestaustralien*. Heidelberg: heiBOOKS.

Wergin, Carsten, and Edwin Lee Mulligan. 2021. "Listening 'Care-Fully': Acoustemologies of Hope in the Face of Human-Made Environmental Degradation." *Cadernos de Arte e Antropologia* 10 (2): 24–37.

Wergin, Carsten, and Stephen Muecke. 2012. "Questions of Value: Tourism and the Resources Boom." *Australian Humanities Review* 53. Accessed December 22, 2016. http://australianhumanitiesreview.org/2012/11/01/questions-of-value-tourism-and-the-resources-boom/.

White, David. 1993. "Tourism as Economic Development for Native People Living in the Shadow of a Protected Area: A North American Case Study." *Society and Natural Resources* 6 (4): 339–45. doi:10.1080/08941929309380832.

White, Sandra, Peter Williams, and Terry Hood. 1998. "Gearing Up for Aboriginal Tourism Delivery: The Case of First Host." *Journal of Hospitality and Tourism Education* 10 (1): 6–12. doi:10.1080/10963758.1998.10685166.

Whitford, Michelle, Barry Bell, and Mike Watkins. 2001. "Indigenous Tourism Policy in Australia: 25 Years of Rhetoric and Economic Rationalism." *Current Issues in Tourism* 4 (2/4): 151–81. doi:10.1080/13683500108667886.

Whitford, Michelle, and Lisa Ruhanen. 2016. "Indigenous Tourism Research, Past and Present: Where to from Here?" *Journal of Sustainable Tourism* 24 (8–9): 1080–99. doi:10.1080/09669582.2016.1189925.

Williams, Peter W., and J. Kent Stewart. 1997. "Canadian Aboriginal Tourism Development: Assessing Latent Demand from France." *Journal of Tourism Studies* 8 (1): 25–41. doi:10.1177/004728759803700117.

Woodside. 2012. "Browse LNG Development: World Class Whale Research. Fact Sheet." Accessed April 29, 2013. http://www.woodside.com.au/our-business/browse/documents/browse%20fact%20sheet%20whales.pdf.

Zeppel, Heather D. 2006. *Indigenous Ecotourism: Sustainable Development and Management*. Wallingford: CABI.

FURTHER READINGS

Acheson, James M. 2006. "Institutional Failure in Resource Management." *Annual Review of Anthropology* 35 (1): 117–34. doi:10.1146/annurev.anthro.35.081705.123238.

Ballard, Chris, and Glenn Banks. 2003. "Resource Wars: The Anthropology of Mining." *Annual Review of Anthropology* 32 (1): 287–13. doi:10.1146/annurev.anthro.32.061002.093116.

Barad, Karen. 2007. *Meeting the Universe Halfway: Quantum Physics and the Entanglement of Matter and Meaning*. Durham: Duke University Press.

Behrends, Andrea, Stephen Reyna, and Günther Schlee, eds. 2011. *Crude Domination: An Anthropology of Oil*. Oxford: Berghahn Books.

Boltanski, Luc, and Laurent Thévenot. 2006. *On Justification: Economies of Worth*. Princeton: Princeton University Press.

Breitfeld, Juliane. Forthcoming. "Images Circulating in a Contested Landscape: Rock Art Networks in the Kimberley Region, Western Australia." PhD Dissertation. Ruprecht-Karls-University Heidelberg.

Carrier, James G., and Paige West, eds. 2009. *Virtualism, Governance and Practice: Vision and Execution in Environmental Conservation*. New York: Berghahn Books.

Carrithers, Michael, Matei Candea, Karen Sykes, Martin Holbraad, and Soumhya Venkatesan. 2010. "Ontology Is Just Another Word for Culture: Motion Tabled at the 2008 Meeting of the Group for Debates in Anthropological Theory, University of Manchester." *Critique of Anthropology* 30 (2): 152–200. doi:10.1177/0308275x09364070.

Carson, Dean, Doris Carson, and Andrew Taylor. 2013. "Indigenous Long Grassers: Itinerants or Problem Tourists?" *Annals of Tourism Research* 42: 1–21. doi:10.1016/j.annals.2013.01.009.

Clark, Ian D. 2002a. "Rock Art Sites in Victoria, Australia: A Management History Framework." *Tourism Management* 23 (5): 455–64. doi:10.1016/s0261-5177(02)00011-0.

Clark, Ian D. 2002b. "The Ebb and Flow of Tourism at Lal Lal Falls, Victoria: A Tourism History of a Sacred Aboriginal Site." *Australian Aboriginal Studies* 2: 45–53.

Clark, Ian D., and Eva McRae-Williams. 2013. "Tourist Visitation to Ebenezer Aboriginal Mission Station, Victoria, Australia, 1859–1904: A Case Study." *Tourism Culture and Communication* 13 (2): 113–23. doi:10.3727/109830413x13848886455272.

de Castro, Eduardo Viveiros. 2015. *The Relative Native: Essays on Indigenous Conceptual Worlds*. Chicago: HAU Books.

de la Cadena, Marisol. 2015. *Earth Beings: Ecologies of Practice across Andean Worlds*. Durham: Duke University Press.

DeLanda, Manuel. 2006. *A New Philosophy of Society: Assemblage Theory and Social Complexity*. London and New York: Continuum.

Farmer, Paul. 2004. "An Anthropology of Structural Violence." *Current Anthropology* 45 (3): 305–25. doi:10.1086/382250.

Haraway, Donna. 2016. *Staying with the Trouble. Making Kin in the Chthulucene*. Durham: Duke University Press.

Healy, Chris. 1999. "White Feet and Black Trails: Travelling Cultures at the Lurujarri Trail." *Postcolonial Studies* 2 (1): 55–73. doi:10.1080/13688799989896.

Higgins-Desbiolles, Freya, George Trevorrow, and Syd Sparrow. 2014. "The Coorong Wilderness Lodge: A Case Study of Planning Failures in Indigenous Tourism." *Tourism Management* 44: 46–57. doi:10.1016/j.tourman.2014.02.003.

Ho, Henry W., and Shameem Ali. 2013. "Understanding Negative Visitor Experiences at Indigenous Cultural Tourism Venues: Marketing and Operational Implications." *Journal of Marketing Development and Competitiveness* 7 (2): 138–45.

Kirsch, Stuart. 2002. "Anthropology and Advocacy: A Case Study of the Campaign against the Ok Tedi Mine." *Critique of Anthropology* 22 (2): 175–200. doi:10.1177/03075x02022002851.

Kirsch, Stuart. 2006. *Reverse Anthropology: Indigenous Analysis of Social and Environmental Relations in New Guinea*. Stanford: Stanford University Press.

Kirsch, Stuart. 2007. "Indigenous Movements and the Risks of Counterglobalization: Tracking the Campaign against Papua New Guinea's Ok Tedi Mine." *American Ethnologist* 34 (2): 303–21. doi:10.1525/ae.2007.34.2.303.

Latour, Bruno. 1998. "To Modernise or to Ecologise, That is the Question." In *Remaking Reality: Nature at the Millennium*, edited by Noel Castree and Bruce Willems-Braun, 221–42. London and New York: Routledge.

Latour, Bruno. 2013b. "Biography of an Inquiry: On a Book about Modes of Existence." *Social Studies of Science* 43 (2): 287–301. doi:10.1177/0306312712470751.

Latour, Bruno. 2018. *Down to Earth: Politics in the New Climatic Regime*. Cambridge: Polity Press.

Li, Tania Murray. 2000. "Articulating Indigenous Identity in Indonesia: Resource Politics and the Tribal Slot." *Comparative Studies in Society and History* 42 (1): 149–79. doi:10.1017/s0010417500002632.

Malm, Andreas. 2016. *Fossil Capital: The Rise of Steam Power and the Roots of Global Warming*. London: Verso.

Mantz, Jeffrey W. 2008. "Improvisational Economies: Coltan Production in the Eastern Congo." *Social Anthropology* 16 (1): 34–50. doi:10.1111/j.1469-8676.2008.00035.x.

Morton, Adam. 2018. "Climate Change: Tourism is Australia's Least Prepared Industry, Report Says." *The Guardian*. Accessed May 3, 2018. https://www.theguardian.com/environment/2018/feb/08/climate-change-tourism-is-australias-least-prepared-industry-report-says#img-1.

Murrey, Amber. 2016. "Antipode: A Radical Journal of Geography." In *Review of Subterranean Estates: Life Worlds of Oil and Gas*, edited by Hannah Appel, Arthur Mason, and Michael Watts. Accessed November 30, 2016. https://radicalantipode.files.wordpress.com/2016/03/book-review_murrey-on-appel-et-al.pdf.

Orlove, Ben, and Steven C. Caton. 2010. "Water Sustainability: Anthropological Approaches and Prospects." *Annual Review of Anthropology* 39 (1): 401–15. doi:10.1146/annurev.anthro.012809.105045.

Pannell, Sandra. 1993. "'Circulating Commodities': Reflections on the Movement and Meaning of Shells and Stories in North Australia and Eastern Indonesia." *Oceania* 64 (1): 57–76. doi:10.1002/j.1834-4461.1993.tb02447.x.

Petri, Helmut. 1954. *Sterbende Welt in Nordwestaustralien*. Braunschweig: Albert Limbach.

Pomering, Alan. 2013. "Indigenous Identity in the Nation Brand: Tension and Inconsistency in a Nation's Tourism Advertising Campaigns." *Corporate Reputation Review* 16 (1): 66–79. doi:10.1057/crr.2012.26.

Pomering, Alan, and Leanne White. 2011. "The Portrayal of Indigenous Identity in Australian Tourism Brand Advertising: Engendering an Image of Extraordinary Reality or Staged Authenticity?" *Place Branding and Public Diplomacy* 7 (3): 165–74. doi:10.1057/pb.2011.19.

Povinelli, Elizabeth A., Mat Coleman, and Kathryn Yusoff. 2017. "An Interview with Elizabeth Povinelli: Geontopower, Biopolitics and the Anthropocene." *Theory, Culture & Society* 34 (2–3): 169–85.

Rowe, David. 2012. "The Bid, the Lead-Up, the Event and the Legacy: Global Cultural Politics and Hosting the Olympics." *The British Journal of Sociology* 63 (2): 285–305. doi:10.1111/j.1468-4446.2012.01410.x.

Savinovic, Ana, Sangkyun Kim, and Philip Long. 2012. "Audience Members' Motivation, Satisfaction, and Intention to Re-Visit an Ethnic Minority Cultural Festival." *Journal of Travel & Tourism Marketing* 29 (7): 682–94. doi:10.1080/10548408.2012.720154.

Sawyer, Suzana. 2004. *Crude Chronicles: Indigenous Politics, Multinational Oil, and Neoliberalism in Ecuador*. Durham: Duke University Press.

Scherrer, Pascal, Amanda J. Smith, Martin Randall, and Ross Dowling. 2011. "Environmental and Cultural Implications of Visitor Access in the Kimberley Region, Australia." *Australian Geographer* 42 (3): 257–71.

Smith, Sherry L., and Brian Frehner. 2010. *Indians and Energy: Exploitation and Opportunity in the American Southwest*. Santa Fe: School for Advanced Research Press.

Spivak, Gayatri Chakravorty. 2003. *Death of a Discipline*. New York: Columbia University Press.

Stengers, Isabelle. 2010. *Cosmopolitics*. Minneapolis: University of Minnesota Press.

Tsing, Anna L. 2015b. "A Feminist Approach to the Anthropocene: Earth Stalked by Man." Accessed December 2, 2016. https://vimeo.com/149475243.

UNWTO (United Nations World Tourism Organization). 2012. *Compendium of Tourism Statistics: Data 2006-2010*. Madrid: World Tourism Organization.

Zimmermann, Erich. 1933. *World Resources and Industries: A Functional Appraisal of the Availability of Agricultural and Industrial Resources*. New York: Harper and Row.

Index

ABC Four Corners, 50
Aboriginal and Torres Strait Islander Studies (AIATSIS), 19
Aboriginal Community Development Policy (ACDP), 40
Aboriginal Heritage Act 1972 (WA), 196
agential realism, 63, 79
Altman, Jon, 139–41
anthropocene (the age of), 56–60, 88, 117, 161
archaeology, 22, 180, 190–94
Asia (cultural presence), 2, 37, 59, 61, 77, 150
assemblage (theory), 100, 203. *See also* resource
Australia: Aboriginal Australia, 1, 55, 113; Australian Dreaming, 2, 4, 53, 62, 124, 127, 203; "Multicultural Australia," 5–6, 13, 55, 69, 86–89, 93, 136, 202; Transcultural Australia, 56, 93
authenticity, 69, 75, 114, 128
Azuki (Japanese restaurant), 37

Bagshaw, Geoffrey, 158
Bánman/Mában (healer), 183
Barad, Karen, 79
Bardi, 153
Barnett, Colin, 34, 50
Barred Creek, 135, 138
Bates, Daisy, 159
Beagle Bay, 25, 39–41, 86, 88, 150–51
Bidjadanga, 47, 150
bilbies, 3, 103; bilby burrows, 165; Greater Bilby, 161
Bindinyanygun/Yellow River, 47, 134
Bindunbur, xi, 116, 147
Bird Rose, Deborah, 90
blackbirding, 150, 157, 176n1
"Black Tuesday," 34–35, 138–39, 163
blood lineage, 3, 148, 154
Bradshaw and Fry (report), 9, 189
Broome, xii, 4, 5, 36; Common Gate, 39, 157; community, 39, 41, 44–46; Creole culture, 77, 152; Crocodile Park, 33, 103; hospital, 109; Lotteries House, 24, 125, 161–71; Old Hostel, 27, 39, 42, 119, 130; Old Town, 32; Our Lady Queen of Peace Cathedral, 32, 43; tourism economy, 76, 103, 125; Visitor Centre, 108
Broome Chamber of Commerce and Industry (BCCI), 124
Broome Cup, 41
Broome Tourism Office (BTO), 124

Brown, Bob, 188
Brown, Jeanné, 46, 118, 129, 134
Browse Basin, 2
Bugarrigarra, 6, 38, 61, 85, 91, 101, 115, 123, 126, 159, 181
burial sites, 3, 9, 73, 180–81, 189, 191–92,195
Butler, John, 53

Cable Beach, 4, 28, 32–33, 38, 43, 58, 103–5, 118–19, 135, 162
Cane, Scott, 159
Cape Leveque Road, 25, 47
Chi, Jimmy, 37
childcare, 31, 42; Circle House, 24
citizen science, 24, 142, 147–49, 161–70, 174–76
co-becoming, 63–66, 79
collaboration(s), 28, 36, 79, 85, 95, 130–33, 189, 202–3; collaborator(s), 3–6, 22–24, 63, 142, 194, 201
compensation package, 92, 132, 153, 187
concert(s), 24, 31–33, 53, 162–63
Conservation Volunteer Australia, 166
consultant firms, 148, 160, 166–70; RPS Consultants, 170
"contact situation," 58, 85–92
corroboree(s), 39, 49, 117, 143
Council for Aboriginal Reconciliation, 21
country, 2–6; cultivated country, 111–14, 118–19, 124, 133, 150, 197; healthy country, 90; living country, 6–9, 18–23, 38–39, 61, 97, 110–19, 126–33, 160–70; nourishing (country), 90–91; respect for country, 134. *See also* living heritage
Courthouse Markets, 41–47, 125
cultural diversity, 22, 43
custodians, 3–8, 13, 34–35, 48, 68, 70, 90, 111–16, 147–51, 189

Dabber Dabber Goon, 115
Dampier Peninsula, 25, 116, 158–62, 176n2
decolonization, 58, 64, 89, 160, 203
Deepwater Horizon, 96
Deleuze, Gilles, 100
Development, 5, 62, 68, 72, 202–3; community, 40, 121n9; industrial, 8, 58, 67, 78, 101, 138; sustainable, 23, 168. *See also* Indigenous tourism
dinosaur trackways, 3, 73, 115, 126. *See also Marala*
diplomacy, 18
Dodson, Patrick, 9, 21, 27, 93
"The Dolphin Children," 48
Dora Street, 27, 39–44, 119, 130
Dreaming, xii, 3, 6, 123, 138, 184; Australian Dreaming, 53, 62–63, 127–33, 201–4. *See also* opportunity
dualism(s), 6–8, 31, 35, 62–64, 79–80, 82n3, 88–92. *See also* four pillars of settler-colonialism
dugongs, 161
Dumont, Louis, 175
dunes, 9, 43, 106, 135, 189–95
Dutch disease, 110

ecocide, 117
eco-feminism, 62
Edgar, Jimmy, 35
entanglement, 3, 6, 56, 62, 72, 78, 85, 90–93, 100, 111–12, 134–35, 141, 197. *See also* multispecies; transecology
Environmental Humanities, 7, 22, 55–57, 79
Environmental Protection Authority (EPA), 161–62, 168–77
Environs Kimberley, 24, 27, 34, 44, 49, 161; "ethical clearance," 17–22, 31, 34; to give back to the community, 180. *See also* Pritchard, Martin
ethnography, 5, 7, 17, 53, 56–63, 77, 85; ethnographic writing, 179–83
European Association of Social Anthropologists (EASA), 19
"evidence from the grave," 156–57
extinction, 64–66, 153

"fair go," 12n1, 69, 174–75, 202
families, 24–30, 44–48, 88, 127, 137, 164, 179; "Broome families," 30–32,

162; Hunter family (The Hunters), 10, 138–40; Roe family (The Roes), 10, 42, 113, 138–40, 156–57
Federal Court Justice Anthony North, xi, 116, 150
Federal Court of Australia, 153
"ferals," 149
fireplace, 147, 192
Fitzroy River, 50
fly-in, fly-out (FIFO), 78–79, 102, 121n9, 125
Fong, Kevin, 39
the four pillars of settler-colonialism, 64, 85, 154, 157
Fox, D. C., 183
fracking, 2, 80, 90, 111
Franklin dam, 51–53
"frozen in time," 91, 116, 129, 160

Gandhi, Mahatma (principles), 51, 186, 189
Gantheaume Point, 46–47, 189
German Association of Social and Cultural Anthropology (GASCA), 19
Global South, 89
Goolarabooloo, 8–12, 20, 31, 46–48, 110–13, 137–41, 148–68, 196–99
Goolarri Media, 39, 44, 125
grassroots, 27
Greenpeace, 189
Greens Party (Australian), 188
"grey nomads," 4
grinding stones, 147, 191–94
Guattari, Felix, 100
Gulf of Mexico, 96

Haraway, Donna, 64
Harvard Model, 126
heritage, 4–7, 179–89; heritage sites, 69, 76, 180; living heritage, 3–8, 18, 29, 110–20, 147, 149, 167–68, 194–99; mainstream heritage discourse, 114–16; transcultural heritage, 2, 13n7, 57. *See also* living country
Hoogland, Frans, 8–10, 37, 46, 111–13, 131, 184. *See also Goolarabooloo*
hope, xi, 90; biocultural hope, 53, 66, 97; "radical hope," 17, 22, 199. *See also* otherwise
Hostile Environment Services (The Hostiles), 26–27, 186
Hunter, Richard, 49–51, 131, 134, 141, 157, 191. *See also Goolarabooloo*
hybridity, 8

Indian Ocean, 1–2, 7, 25, 28, 37–38
Indigeneity, 3, 6, 36, 71–72, 78, 100–101, 203
Indigenous: communities, xii, 2, 11, 25; knowledge, 12, 22, 97, 117, 152, 166–75; people, xii, 36, 41, 54n2, 69–75, 78, 90–94, 103, 130, 139, 195; renaissance, 77; tourism, 10, 25, 67–80, 101–3; world (view)s, 16, 85, 90, 100–3, 180–82, 203
Indigenous Policy and Dialogue Research Unit, 21
informed consent, 19, 26
interviews, 22–24, *See also* ethnography; methodology

Jabirr Jabirr/Ngumbarl, xii, 8, 11, 48, 115–16, 148–59
Jandamarra, 37
Japan, 2, 37–43, 150
jila (waterhole), 189–90

Karajarri Native Title determination, 185
kartiya, 131, 163, 168
Keating, Paul (former Prime Minister), 40
Kennedy Hill, 40
"killer" (bullock), 140–41
Kimberley Economic Forum, 35, 58, 125
Kimberley Land Council (KLC), 9, 23, 50, 73–74
Kimberley (region), 2–8, 55–60, 67–80
Kimberley Training Institute (KTI), 75, 86–94
kinship, 3, 85, 92, 137, 183
knots, 27–30, 41. *See also* methodology
knowledge (shared), 3, 133, 141, 163–68; knowledge production,

147–48, 179; Traditional Ecological Knowledge (TEK), 165
Kolig, Erich, 117, 183
Kunkel, Benjamin, 64

"labor," 132, 135, 141
Labor Party (Australian), 170, 188
land, 3; land rights, 59, 75, 85, 143, 156. *See also* country
landscape, 4, 25, 36, 50, 55–56, 112; landscape architecture, 129–31. *See also* transecology
Langton, Marcia, 93
Language, 30; lay language, 19, 151, 179; people-country-language nexus, 155; Western language, 92, 149
"last frontier," 2, 28, 55, 69, 80, 181, 197
Latour, Bruno, xii, 18, 100
Lear, Jonathan, 17
learning, 35, 71, 79, 113, 127–43
linear time, 64, 52, 85, 152–54. *See also* the four pillars of settler-colonialism
liquefied natural gas (LNG) facility, xi, 2, 4, 27, 31
lived experience, 16–17, 64–65, 111, 127–45. *See also* entanglement
liyan, 6, 123, 134, 143, 150–51, 202
local local(s), 32
lock-on, 52, 198
Lombadina, 25, 39, 150
lore, 2–5, 11–12, 13n2, 49, 57–61, 91, 148–76, 182–87
Lowenthal, David, 5
Lurujarri Heritage Trail (the Trail), xi, 3, 7–12, 38, 46, 127–44, 163, 184, 197

Mabo, 40, 85, 91, 130, 137. *See also* Native Title
MacCannell, Dean, 68
MacRobertson Miller Airline (MMA), 107
Maher, Chris, 33
maja (*lore* boss), 112, 154, 157
Mamabulanjin (Aboriginal cooperation), 39–44, 125, 150
Manari Road (corner), 24–27, 34–35, 138–39
Marala (The Emu Man), 115, 126
market economics, 1, 12, 63, 70, 86, 89, 96–100, 123–24, 175; economic opportunity, 123, 133, 201; "hybrid economy," 140–42; "Land of Opportunity," 1, 118; progress, 36, 60–62, 75, 98, 185. *See also* "opportunity dreaming"
materialities, 63, 98, 100, 147
Matso's restaurant, 43, 142–43
McAlpine, Lord Alistair, 9, 103–8, 118, 127
medical services, 42, 109
megalosauropus broomensis, 115
methodology, 18–20, *See also* ethnography; interviews; participant observation
The Mexicans (band), xii, 108
middens, 9, 15, 136, 190–92
Middleton, Louise, 164, 198
Minas Gerais, 96
mineral sands, 130, 212, 220
mining, 8–10, 61, 78, 80–83, 101–3, 130–33
mission(arie)s, 39, 93, 152, 183; Pallottines, 151
monsoonal vine thickets, 3, 164, 202
more-than-human, 3, 62–64, 79, 98, 147, 194
Moreton-Robinson, Aileen, 90
"Mother's Day Protest," 30, 99
Mr Roe, xi, 13n5, 20–21, 49–52, 137, 163, 168, 191, 198
Muecke, Stephen, xii, 36–39, 42–43, 93, 111, 160, 191
multispecies, 56, 64–66, 147, 201
Myers, Fred, 113
myth-consciousness, 91, 137, 183–85

naming, 64, 85, 93, 96, 157. *See also* the four pillars of settler-colonialism
Nangan, Joseph "Butcher Joe", 39
National Aboriginal and Torres Strait Islander Tourism Strategy, 75

Native Title, 20, 48, 65–66, 75, 85, 92–93, 114–18, 132–33, 147–57
Ngunungurrukun (*Millibinyarri*/ Coconut Wells), 28, 46–48, 118–19, 131, 163
Nirrumbuk, 41
No Gas campaign (No Gas), 27, 32–35, 51–52, 125, 161, 165
Non-Violent Direct Action (NVDA), 51–52
Northern Tradition, 7, 11, 154
"Nowhere else but here," 15, 30–35, 125, 188
Nyamba Buru Yawuru, 20
Nyigina, 20, 39, 153–54, 158–59
Nyul Nyul, 41, 115–16, 148–53

Old Parliament House, 29
One Arm Point, 9, 25, 40, 47
ontological turn, 89; ontological differences, 91, 139, 160, 183
Operation Kimberley *Miinimbi*, 188
"opportunity dreaming," 2, 6, 80, 118, 124, 136–38, 151, 175, 184–85, 203. *See also* economic opportunity
otherwise, 12, 56–57, 65–66, 90, 96–97, 197, 202–4. *See also* hope
ownership, 49, 64, 75, 85, 90, 148, 155–60. *See also* the four pillars of settler-colonialism

Parke, Melissa, 170, 188
participant observation, 22–24
pearling, 41, 69, 103–11, 176n1; pearl divers, 2, 43, 150; pearl lugger, 106–8
Pedersen, Howard, 37
Pender Bay, 41
Peninsula Bombers (Australian football team), 94
Petri, Helmut, 85, 183
Petri-Odermann, Gisela, 183
Philippines, 43
The Pigram Brothers (The Pigrams), 30–34, 37, 162; *See also* "Nowhere else but here"
Pigram, Stephen, 40, 93, 163
pindan (sand), 16, 40, 104
plant use, 130
Pley, Kees, 40
Plumwood, Val, 62–63
Poelina, Anne, 34, 50, 117
Povinelli, Elisabeth, 17, 112
preservation, 1–10, 69, 75–76, 113–16, 202
Pritchard, Martin, xii, 27, 34, 50–53, 149, 190–91, 198
professional primitivism, 77
protest, 23–34, 49–53, 99, 117, 138–39, 149, 186–90

rayi (spirit), 153–58
reconciliation, 21, 77, 92–95, 144
recreational vehicles (RV), 4, 108
regime of justification, 175
regional elections, 124
remote(ness), 2, 5, 67–72, 109, 128, 150
representationism, 17–18, 63–64, 117, 142, 160, 182–86
Research Executive Agency (REA), 19
resources boom, 2–4, 56, 63, 100, 124, 133; resource exploration, 1, 69, 78, 97–101, 179, 197; "resourceness," 96–101, 124, 133
ritual syncretism, 159
Roach, Archie, 93
Roe, Paddy OAM (the Old Man), 8–10, 39–48, 111–13, 130–39, 153–60, 184, 189–91
Roe, Phillip (Phil), 43, 142–43, 163–68, 191–94
Roe, Theresa, 8, 42–43, 48, 153–58
Roebuck Bay, 36
Royal Melbourne Institute of Technology (RMIT), 15, 129–31, 135, 162
Rubibi Claim Group, 20

Sahlins, Marshall, 49
Sansom, Basil, 203
Save the Kimberley, 161
saxophone, 31, 44, 143
Sea Shepherd, 188–89

settler-colonialism, 12, 36, 61–65, 85, 91–92, 152–57, 185
shallow generational memory, 159
share, 5, 62–69, 113, 118; shared experience, 138, 164. *See also* the Trail; values
Shinju Matsuri, 41–45, 125
Sinatra, Jim, 129–31
skull, 29
social anthropology, 7, 18, 19, 54, 64, 68, 98, 131, 182; anthropologists, 9, 17, 23–24, 31, 39, 54n2, 64, 85, 90–91, 113, 139, 179–87
Social Policy Research Centre (SPRC), xiii, 22
Song Cycle, 7–10, 47–49, 112, 115, 118, 130, 155–58, 184
sovereignty, 66, 90
"speak well," 18–19
"Staircase to the moon," 41
Stanner, William Edward Hanley, 90–91
Steve Irwin (ship), 188, 194
Stolen Generation, 86, 150–52, 186
succession, 153–54, 159, 176n2
sustainability, 6, 10, 23, 76
Sutton, Peter, 155, 159

terra nullius, 154
Terrex, 9
tidal movements, 129, 135, 173
Todd, Zoe, 89
totemism, 91
tourism (in the Kimberley), 55, 67–83; collaborative tourism, 73; conventional tourist experiences, 136, 141; Indigenous tourism, 10, 22–25, 66–80, 101–10, 127–28, 188; reconciliation tourism, 77; tourism imaginaries, 55; "tourist dollar," 105–7, 111; "tourist season," 4, 36
Traditional Owner (TO), 48–49, 73–74, 103, 166–67, 174
transculturality, 12, 55
transecology, 12–13, 55, 126, 133, 144, 167, 175, 186, 194–97
transitional zone, 158–60
translation, 22, 58, 63, 127, 175, 180–82, 197; translatability, 22, 186
Turnbull, Malcolm (former Prime Minister), 5, 13, 55, 202
turtles, 3, 141, 147, 161–70, 201; Hawksbill turtle, 161; turtle bones, 147, 190–92; turtle nesting, 3, 166

The University of New South Wales (UNSW), 21

value(s), 2–10, 53, 56–61, 64, 92–95, 100–3, 113–4, 163–64, 202; existential value, 135; intrinsic values (of a place), 56, 68, 114, 147; regimes of value, 31, 34–35, 57, 65, 101, 114, 124

Walmadany/James Price Point, 1–3, 7, 10, 29–30, 43, 65, 132–33, 169–74, 190, 194–99; concert, 53; protest camp, 26, 34, 186–87; Tent Embassy, 29; warrior, 192
Weaver, David, 78
Welcome to Country, 35
The West Australian (newspaper), 4
Western Australia (WA), 2, 68, 116
whale migration, 3, 169–73; humpback whales, 169–73, 188–89; Kimberley Community Whale Research Project, 170
White Australia Policy, 150
whitefella culture, 48, 72, 134, 168, 182
"white magic," 131, 136, 147, 167, 180–81, 186, 196
Willie Creek Pearl Farm, 25, 49, 69, 107, 111, 118, 125, 139
Woodside Petroleum Ltd., 10–11, 26, 34, 116, 132, 163–73, 198–99
world-making, 16, 63, 89, 95–96, 197, 202. *See also* the Trail
world (view)s, 5–6, 12, 16–22, 57–65, 89–96, 124–29, 135–42, 180–88

Yalagarr (childcare centre), 42
Yáobida (the speaker), 183–85, 194
Yawuru (PBC), 20–21
Yu, Peter, 40, 93

Zanders (restaurant), 38

About the Author

Carsten Wergin is associate professor of anthropology at Ruprecht-Karls-University Heidelberg (Germany). He is the author of various books, edited volumes, and articles in the fields of heritage, culture, and ecology and serves as President of the German Association for Australian Studies (GASt) and on the executive committee of the Association for Critical Heritage Studies (ACHS).